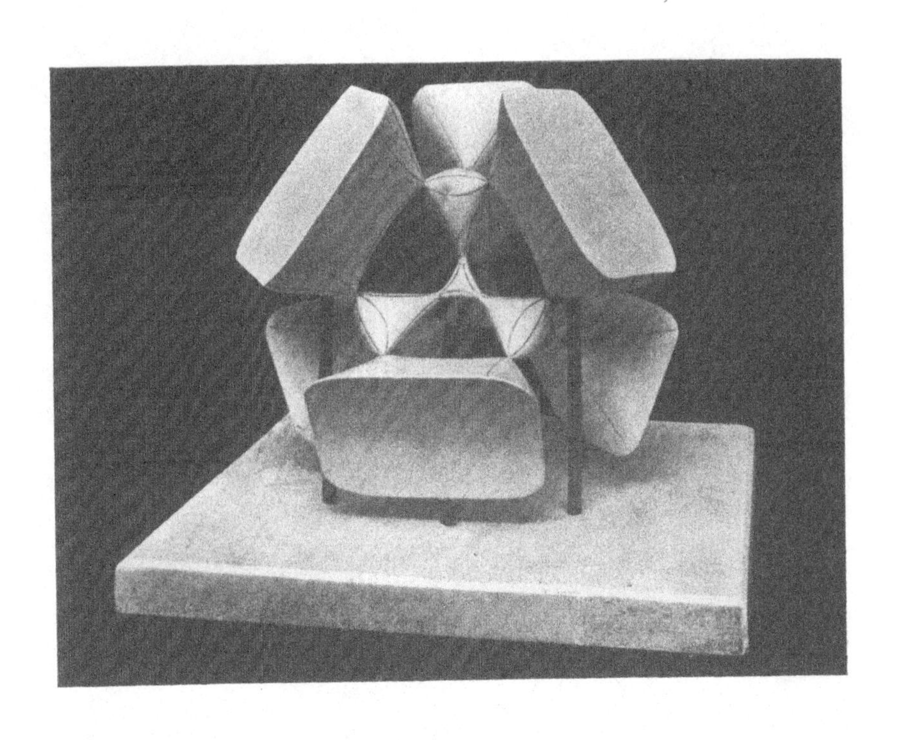

KUMMER'S QUARTIC SURFACE

KUMMER'S QUARTIC SURFACE

BY

R. W. H. T. HUDSON, M.A., D.Sc.

Late Fellow of St John's College, Cambridge, and Lecturer in Mathematics
at the University of Liverpool.

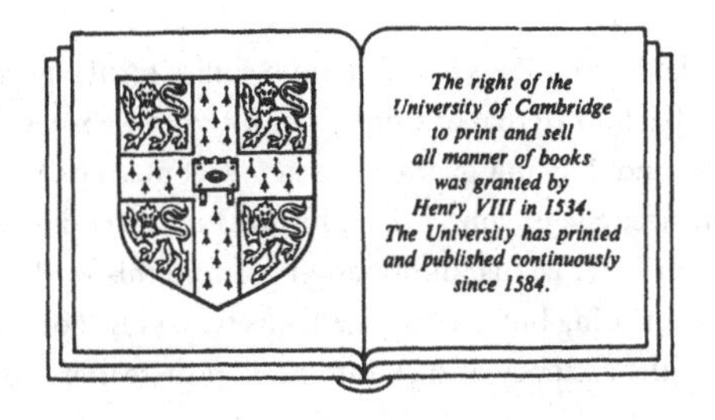
The right of the
University of Cambridge
to print and sell
all manner of books
was granted by
Henry VIII in 1534.
The University has printed
and published continuously
since 1584.

CAMBRIDGE UNIVERSITY PRESS

Cambridge

New York Port Chester

Melbourne Sydney

CAMBRIDGE UNIVERSITY PRESS
Cambridge, New York, Melbourne, Madrid, Cape Town,
Singapore, São Paulo, Delhi, Tokyo, Mexico City

Cambridge University Press
The Edinburgh Building, Cambridge CB2 8RU, UK

Published in the United States of America by
Cambridge University Press, New York

www.cambridge.org
Information on this title: www.cambridge.org/9780521397902

First published 1905
Reissued as a paperback with a Foreword by W. Barth
in the Cambridge Mathematical Library series 1990

A catalogue record for this publication is available from the British Library

Library of Congress Cataloguing in Publication data

ISBN 978-0-521-39790-2 Paperback

CONTENTS.

CHAPTER IV.

LINE GEOMETRY.

CHAPTER V.

THE QUADRATIC COMPLEX AND CONGRUENCE.

CHAPTER VI.

PLÜCKER'S COMPLEX SURFACE.

CHAPTER VII.

SETS OF NODES.

CHAPTER VIII.

EQUATIONS OF KUMMER'S SURFACE.

CHAPTER IX.

SPECIAL FORMS OF KUMMER'S SURFACE.

CHAPTER X.

THE WAVE SURFACE.

CHAPTER XI.

REALITY AND TOPOLOGY.

CHAPTER XII.

GEOMETRY OF FOUR DIMENSIONS.

CHAPTER XIII.

ALGEBRAIC CURVES ON THE SURFACE.

CHAPTER XIV.

CURVES OF DIFFERENT ORDERS.

CHAPTER XV.

WEDDLE'S SURFACE.

CHAPTER XVI.

THETA FUNCTIONS.

CHAPTER XVII.

APPLICATIONS OF ABEL'S THEOREM.

CHAPTER XVIII.

SINGULAR KUMMER SURFACES.

FOREWORD

In 1916 there appeared, a decade after Hudson's book, Jessop's book on quartic surfaces (Jessop, 1916). In his preface Jessop states 'The admirable work written by the late R. W. H. T. Hudson, entitled *Kummer's Quartic Surface*, renders unnecessary the inclusion of this subject', i.e. the inclusion of Kummer surfaces in his own book. Now, almost a century later, and after so many deep changes of methods, style and aims in geometry, Hudson's work stands as admirably as when it impressed the experienced author, Jessop.

The theory of Kummer surfaces includes so many beautiful aspects of geometry that A. Weil used the first letter of its inventor's name, when he introduced the notion $K3$ for the class of surfaces to which Kummer surfaces belong. To some extent the beauty and importance of the subject explains the attraction of Hudson's book. To a larger extent, however, the work impresses by its masterly presentation. Hudson starts at the beginning - nothing is assumed to be known except for the elements of geometry in space. Everything else is explained by the author. Of course this cannot mean he proves everything; often he borrows results from other sources. And naturally he does not always give exact definitions of his mathematical tools, because ideas like 'group actions' or 'varieties' were developed in full precision only later. But he does have the ability to explain!

One of the outcomes of Bourbaki's efforts to modernize our science is the experience, probably not wanted by Bourbaki, that explaining is not the same as defining and proving something. It is the art of presenting material attractively, vigorously, following a path that seems natural to a reader not yet familiar with the final aim. Hudson's book stands out above all others I know by the masterful way in which he, though then not yet thirty, presented the many-faceted material.

Today, the formal theory of surfaces has reached a certain stage of completeness. Of course there are important open questions. But it seems that major efforts concentrate on solving these concrete questions and not on developing further the formal theory. (The two aspects of mathematics are, admittedly, hard to separate.) Some of these questions are already touched on very directly, but elegantly, in Hudson's book: the classification of quartic surfaces, the description of moduli spaces for abelian surfaces with level structures, with distinguished subgroups, the automorphism group of a Kummer surface, ... So I believe a reissue of this book now, when the original is hard to get on the market, will be of use to many doing active research.

In 1972 I saw this book for the first time when checking through the geometry section of Leiden's Mathematics library – where books are arranged by fields and not by the cruel regime of the alphabet only. Since then it has become my favourite book in Mathematics. Several times I have tried to order a copy from catalogues of old books, but never was fast enough. So it was a great pleasure for me that Cambridge University Press offered me the unique opportunity of taking part in this reissue. The book speaks for itself. I want to describe below, in the language of today, the contents of its eighteen chapters, hoping that this will be of use to those readers who are educated in the modern way.

I. KUMMER'S CONFIGURATION

Kummer's configuration of sixteen points and phases in three-space is most easily understood after describing its group of symmetries. In some fixed coordinate system this group is generated by the operations of changing the sign of two coordinates and permuting the four coordinates in pairs (see § 4). It is one of the simplest examples of what is nowadays called the Heisenberg group (see Mumford (1966)). The configuration is obtained from a general plane (or dually a point) as the sixteen transforms of this plane. There will be an orbit of sixteen points under this group such that each configuration plane contains six of these points and, conversely, each point in the orbit lies on six planes: Kummer's 16_6, see § 3. The six points in any plane are in special position in the sense that they lie on a conic (§ 7).

When describing this geometry, Hudson simultaneously introduces the notion of a group. Since he refers to Burnside's book for the definition of a group, and since he obviously has problems when explaining the group operation, we may safely conclude that in about 1900 a group was some complicated concept, not known to the general mathematical public.

In § 1 Hudson describes a pencil of quartic surfaces which are invariant under the group just mentioned, even under some bigger group. This pencil contains three tetrahedrons as surfaces decomposing into four planes. On several later occasions these tetrahedrons appear again, but not the general surfaces in the pencil. This is an example for elementary geometry leading immediately to nontrivial surfaces. In fact these surfaces are birational to Kummer surfaces again. (They are related to self-products $E \times E$ of elliptic curves. The reader may consider this an exercise and use chapter II in Jessop (1916) as a hint.)

II. THE QUARTIC SURFACE

In this chapter Hudson develops the elementary geometry of quartic surfaces in three-space carrying sixteen nodes, which is due to Kummer. Beginning with the definition that a surface is singular where it 'ceases to be approximately flat' he repeats Kummer's proof that its nodes and tropes (= planes touching the surface along a conic) form a 16_6. In chapter I he stated - without convincing proof - that each 16_6 was of the form considered in this chapter.

Hudson proceeds to study the quartic equation of the surface by projecting it from one node. He introduces six parameters $k_1, \ldots, k_6$ for six points on a conic. The conic is the basis for the tangent cone at the fixed node, and the six points give the points of contact on this conic of the six tropes. He uses two copies of the square root

$$\sqrt{(u-k_1)(u-k_2)\ldots(u-k_6)}$$

to parametrize the surface. This is nothing but the uniformization of Kummer's surface by the symmetric product of a hyperelliptic (genus 2-)curve, thus by the jacobian of this curve. The numbers $k_1, \ldots, k_6$ are the coordinates of the six branch points of this curve. § 11 is quite interesting for modern readers. Up to here not one word has been used to say whether Hudson considers the complex surface or its real points only. Here he describes the 'shape' of the surface, of course the real one. He accidentally remarks that the equation of the surface is the linear combination of a tetrahedron and the square of a quadric. He then specializes to the most symmetric case - which, by the way, was exhibited 30 years earlier in a series of plaster models by Kummer - and tries to explain the shape of it. The beautifully symmetric drawings on pages 21 and 22 should be considered a challenge to each owner of a personal computer able to do graphics.

III. THE ORTHOGONAL MATRIX OF LINEAR FORMS

It is a deep satisfaction for us newly educated geometers to have a look at §§ 12–14. Hudson struggles with orthogonal matrices, or rather he tries to present them to his readers in a harmless way. We stop smiling, however, when we see what Hudson does with these orthogonal matrices: he obtains linear and quadratic relations between the sixteen trope planes of the Kummer surface and he explicitly represents sets of eight associated points amongst the sixteen nodes. (Such a set is the intersection of three quadrics.) The ten fundamental quadrics referred to in § 17 are the ten eigenfunctions of the Heisenberg group mentioned above for its operation on the (ten-dimensional) space of quadratic functions on projective three-space. Just as on this space of quadratic polynomials, the Heisenberg group acts commutatively on the six-dimensional space of alternating functions in the coordinates, i.e. on the five-space in which the Plücker quadric lies. Consequently there are six invariant linear forms in the Plücker coordinates, i.e. six invariant quadratic complexes described in § 18.

IV. LINE GEOMETRY

Line geometry is the geometry of lines in three-space, i.e. the geometry of the Plücker quadric. A linear (line-) complex is a hyperplane section of this Plücker quadric. It is defined by an alternating bilinear form on the four-dimensional vector space underlying projective three-space. Orthogonality with respect to this form leads to a symmetric relation between points and planes (null-point to a plane, null-plane for a point) and between lines (two lines are polar if their defining two-dimensional vector spaces are orthogonal to each other). Any first-year student following a course on Linear Algebra knows that, but has probably never heard about apolarity, a kind of commutativity of linear complexes, explained in absolutely elementary geometric terms in § 21.

Apolarity of two linear complexes just means their alternating bilinear forms wedge each other to zero (for these tensors a symmetric bilinear relation). Writing S, T for the correlation defined by these tensors, Hudson writes $ST = TS$ for the apolarity condition, where we would view the tensors S, T as anti-selfadjoint maps $\mathbb{C}^4 \to (\mathbb{C}^4)^*$ and write $S^{-1}T = T^{-1}S$ for apolarity. Then clearly $S^{-1}T$ is an involutory linear map on $\mathbb{C}^4$. Or we could say, two linear complexes are apolar if their hyperplanes in five-space are polar with respect to the Plücker quadric. Then, for dimensional reasons, there are maximally six mutually apolar linear complexes leading to fifteen involutions on our basic three-space. With great love for details, Hudson shows geometrically (see in particular § 26) that together with the identity they constitute the Heisenberg group again. As a by-product of this digression on line geometry (nowadays part of multilinear algebra) Hudson obtains a 16_6 configuration starting from one of its vertices by taking the six null planes for six apolar complexes and the ten polar planes for ten related quadrics. This is the most basic access to the fundamental $16 = 10 + 6$ decomposition of sixteen objects by parity.

Only in the last three sections of this chapter does Hudson use coordinates to complement his, up-to-then, purely geometric method.

This chapter illustrates how easy it is to approach beautiful geometry. Of course, the language is old-fashioned, but I have never seen a modern exposition of all this lovely geometry.

V. THE QUADRATIC COMPLEX AND CONGRUENCE

A quadratic complex is the variety of lines in three-space parametrized by a quadratic hypersurface in five-space (that projective space where the Plücker quadric lies). In general these two quadratic forms can be simultaneously diagonalized. The corresponding coordinate system defines six hyperplanes on five-space. They are mutually apolar linear complexes, so the situation from the preceding chapter returns.

It was Kummer who, studying quadratic line complexes as models for the system of light rays in an optical instrument, observed a particular surface of focal points for pencils of lines in the complex. This surface, called 'singular surface' in the classical literature and denoted by Φ, is the Kummer surface for the quadratic complex. It has a smooth model, a complete intersection in five-space of three quadrics: the Plücker quadric, the quadratic complex, and a third quadric intimately attached to both of these. The birational relation between this smooth $K3$-surface in five-space and its image with sixteen nodes in three-space is extremely rich in geometric properties, which, though not obvious, are accessible with some effort. A modern account is given in the last fifty pages of the voluminous monograph by Griffiths & Harris (1978).

This discovery of Kummer is without doubt one of the most basic facts of algebraic geometry. It is the starting point of many modern theories.

Compared with any modern exposition, Hudson presents in the seventeen pages of this chapter the relation between the quadratic congruence and its Kummer surface with charming elegance. Of course, he uses extensively his geometric preparations. But most notably he does not bear the burden of the theory of our time.

VI. PLÜCKER'S COMPLEX SURFACE

Plücker's surface is a degenerate form of Kummer's surface when two eigenvalues are coincident, as we would say. What can go wrong if one tries to diagonalize simultaneously two quadrics in five-space, and what happens to the Kummer surface then is presented with great care (e.g. in the table on pp. 230–232 in Jessop (1969)). There are fifty-five cases.

Plücker's surface is a rational quartic with a double line. Its equation and other properties form the contents of this short chapter, at the end of which we again find a section on the shape of the surface. I cannot claim to understand the beautiful picture there. But I am convinced it is worth while to spend some computer time on it.

VII. SETS OF NODES

This is probably the chapter for which a modern presentation would differ most from Hudson's text. Not before chapter XVI does Hudson touch on the uniformization of Kummer's surface by theta functions, i.e. the fact that the Kummer surface is a quotient of an abelian surface. The sixteen nodes, corresponding to the sixteen half-periods on the abelian surface, can be parametrized by the sixteen points in affine four-space over the tiny field $\mathbb{F}_2$ with two elements. The symmetries discussed in this chapter - not without notational difficulty - are governed by the geometry of quadrics over this field $\mathbb{F}_2$. It is fascinating to see, however, how purely combinatorial properties of this sixteen-point set translate into geometry in three-space.

The distinction between Rosenhain and Goepel tetrads is nothing but the distinction between isotropic and non-isotropic planes in this affine space over the finite field.

VIII. EQUATIONS OF KUMMER'S SURFACE

This chapter gives two different quartic equations for Kummer's surface in three-space. These two equations differ by the coordinate system in which they are written, i.e. by the coordinate tetrahedron.

In the first case (§ 53), a coordinate system is used in which the symmetries of the Heisenberg group have the particularly simple form described in chapter I. In modern language this is the choice of a level structure on the 2-torsion group of the covering abelian surface; the parameters A, B, C, D used in the equation are the moduli for abelian surfaces with level-two structure, and the cubic identity satisfied by them is the equation for the moduli space of abelian surfaces with such a level structure.

In the second case a Rosenhain tetrahedron is used. This is the choice in the 2-torsion group $\mathbb{F}_2^4$ of some non-isotropic affine subspace $\mathbb{F}_2^4$. The parameters u, v, w, s (the latter of weight 2) are the moduli of such structures on abelian surfaces. This gives an explicit identification of this moduli space with a weighted projective space.

The explicit description of moduli spaces for abelian varieties, as in these two classical cases, is at the moment a field of quite active research.

IX. SPECIAL FORMS OF KUMMER'S SURFACE

In this chapter geometric properties of special Kummer surfaces are described.

Characterizing the surface by the six numbers $k_1, \ldots, k_6$, the cases are distinguished according to how many involutions leave this sixtuplet invariant, i.e. according to the automorphism group of the hyperelliptic curve: a tetrahedroid is the Kummer surface for the jacobian of a curve admitting an involution other than the hyperelliptic one. Multiple tetrahedroids belong to hyperelliptic curves with even bigger automorphism groups. The list of five cases altogether should be compared with the now well-known classification of hyperelliptic curves according to their group of automorphisms (Geyer, 1974).

X. THE WAVE SURFACE

The wave surface, also called Fresnel's wave surface, is related to the expansion of light rays in certain crystals; for a mathematical discussion see e.g. Knörrer (1986). Up to projective transformations it is the same as the tetrahedroid introduced in chapter IX. However, because of its connection with the physical world, its metrical properties are of particular interest. They are studied in this chapter. Also a parametrization of the surface in terms of two elliptic functions is given. This is possible because the additional involution on the hyperelliptic curve is elliptic, i.e. it comes from a map onto an elliptic curve. Consequently the jacobian surface is isogeneous to a product of two elliptic curves.

XI. REALITY AND TOPOLOGY

Reality problems are some of the most intriguing of algebraic geometry. On the one hand, they are very natural; algebraic geometry started in the 'real world', and when visualizing this kind of geometry, reality is of obvious importance. On the other hand, they are either totally trivial, or extremely complicated. Modern theory only has a modest hold on curves and surfaces in three-space.

A real Kummer surface is one given by an equation with real coefficients. Such a surface need not have real points at all, nor need the sixteen nodes be real.

This chapter classifies all real Kummer surfaces according to their number of real nodes, their number of 'sheets' or 'pieces' into which the real part of the surface decomposes (a notion used intuitively, and probably not too easy to make precise), and according to the elliptic or hyperbolic curvature of these sheets.

XII. GEOMETRY OF FOUR DIMENSIONS

For today's mathematician this chapter is the strangest in the book. The author thought it necessary at this point to introduce his readers to elementary projective geometry in four dimensions. Later on he defines a variety to be a hypersurface in four-space.

In between, he shows how to obtain Kummer's 16_6 configuration by the operations of intersection and projection from the figure of six points in four-space in general position. He applies this to the quartic hypersurface which is the dual of the Segre cubic primal. The latter in symmetric form is given by

$$x_1 + \cdots + x_6 = 0, \qquad x_1^3 + \cdots + x_6^3 = 0.$$

He shows that a hyperplane section of the quartic has fifteen nodes, hence if it is the intersection with a tangent plane it has sixteen nodes and is a Kummer surface. Dually, when projecting the Segre cubic onto three-space from one of its points, the 'enveloping cone is the cone over' such a Kummer surface. In modern terminology: the branch surface is Kummer. Now these results, obtained in a quite elementary way and clearly with a minimum of theory, are of considerable interest in present-day algebraic geometry: phrased somewhat more precisely, they mean that the dual quartic to the Segre primal is a moduli space for abelian surfaces with level-two structure. Each point on this threefold is the modulus point for the abelian cover of its Kummer tangent section. A modern treatment putting this in proper perspective is given, for example, in v. d. Geer (1982).

XIII. ALGEBRAIC CURVES ON THE SURFACE

XIV. CURVES OF DIFFERENT ORDERS

These two chapters give general facts about algebraic curves on the surface and apply them to get something like a classification. Again the arguments given here are elementary, and the resulting equations explicit. Anybody in need of such equations will be very grateful to Hudson. (Such need could arise, for example, if one wanted to draw a computer picture of the surface and of some curves on it.) However, the presentation is really very far from a proper understanding. Modern methods of dealing with symmetric divisors on an abelian surface are so much more powerful. So these two chapters seem to me a part of the book which, from our point of view, should not be placed here, but only after the author shows how to parametrize the surface by theta functions in chapter XVI.

Of course, the advantage of this treatment is that it avoids transcendent theory, so to a large extent is probably independent of the characteristic of the base field.

XV. WEDDLE'S SURFACE

Weddle's surface is another model of a Kummer surface as space quartic, however with only six nodes. It is obtained from the usual model by the linear system of cubics passing through ten 'even' nodes. Hudson first introduces his reader to the theory of birational transformations and linear systems with base points. He touches on the question of which other quartic surfaces are birational images of Kummer surfaces. Weddle's surface is the first interesting case here. He gives geometric properties of the surface and derives its equation.

Again, nowadays we would not try to deal with such questions without the theory of the Picard group and Picard lattice of a Kummer surface. But we should not be conceited: the most natural question in this field, namely to describe the automorphism group of a Kummer surface (= the birational maps into itself) at present is unsolved. We know only a little more than did the classical geometers (see § 120, the last one in this book).

XVI. THETA FUNCTIONS

XVII. APPLICATIONS OF ABEL'S THEOREM

Now, finally, these long-awaited functions are introduced. In the usual, somewhat frightening, formalism, theta-relations are proved, the Kummer surface is parametrized, and the transcendental theory is sketched. Abel's theorem about abelian integrals and divisors of rational functions is applied to derive geometric properties of the Kummer surface. For example, several kinds of tetrahedra inscribed into and circumscribed about the surface are constructed.

These chapters contain a wealth of beautiful geometry. With our modern insight, they are not very astonishing. However, I do not know any other place in the literature where they are collected so nicely.

XVIII. SINGULAR KUMMER SURFACES

Here the classical phrase 'singular' is used to describe a Kummer surface, if the Picard number of its abelian cover is bigger than one. This can happen, for example, if the abelian cover is isogeneous to a product of elliptic curves or, more generally, if it admits a non-trivial endomorphism. It is characterized by relations (linear and quadratic) between the periods of the abelian cover. The surfaces in the moduli space defined by such relations (Humbert surfaces) are at present still of great interest. For example, Hilbert modular surfaces belong to them.

First Hudson describes elliptic Kummer surfaces, i.e. those for which the abelian cover is isogeneous to a product of elliptic curves, and specifies curves on them. For example, he recovers the tetrahedroid there.

Surfaces with 'invariant' five and eight are considered too. Humbert's beautiful descriptions of these surfaces as double covers of the plane ramified over special hexagons circumscribed about a conic are obtained. The abelian cover of a surface with invariant five has real multiplication in the field $\mathbb{Q}\sqrt{5}$, and for invariant eight we have real multiplication in the field $\mathbb{Q}\sqrt{2}$.

W. Barth
Universität Erlangen-Nürnberg

REFERENCES

Geyer, W. D. (1974). *Invarianten binärer Formen*, Springer LNM, 36–39.

Griffiths, Ph. and Harris, J. (1978). *Principles of Algebraic Geometry*, Wiley-Interscience.

Knörrer, H. (1986). *Die Fresnelsche Wellenfläche*, Mathematische Miniaturen 3, Birkhäuser.

Jessop, C. M. (1969). *A Treatise on the Line Complex*, Cambridge (1903), reprinted by Chelsea.

Jessop, C. M. (1916). *Quartic Surface with Singular Points*, Cambridge.

Mumford, D. (1966). On the equations defining abelian varieties. I. *Inv. math.* **1**, 287–354.

v. d. Geer, G. (1982). On the geometry of a Siegel modular threefold. *Math. Ann.* **260**, 317–350.

PREFATORY NOTE.

RONALD WILLIAM HENRY TURNBULL HUDSON would have been twenty-nine years old in July of this year; educated at St Paul's School, London, and at St John's College, Cambridge, he obtained the highest honours in the public examinations of the University, in 1898, 1899, 1900; was elected a Fellow of St John's College in 1900; became a Lecturer in Mathematics at University College, Liverpool, in 1902; was D.Sc. in the University of London in 1903; and died, as the result of a fall while climbing in Wales, in the early autumn of 1904.

This book was then in course of printing, and the writer had himself corrected proofs of the earlier sheets, assisted in this work by Mr T. J. I'A. Bromwich, Professor of Mathematics in Queen's College, Galway, and by Mr H. Bateman, of Trinity College, Cambridge; for the remaining portion Mr Bateman and myself are responsible; we have followed the author's manuscript unaltered throughout; and gratefully acknowledge the care given to the matter by the University Press.

Attentive readers can judge what devotion, what acumen, went to the making of a book of such strength and breadth; a book whose brevity grows upon one with study. To those who knew the writer it will be a reminder of the enthusiasm and brilliance which compelled their admiration, as the loyalty of his nature compelled their regard. A many-sided theory such as that of this volume is generally to be won only by the work of many lives; one who held so firmly the faith that the time is well spent could ill be spared.

H. F. BAKER.

27 *March* 1905.

CHAPTER I.

KUMMER'S CONFIGURATION.

§ 1. DESMIC TETRAHEDRA.

The eight corners of a cube form a very simple configuration; yet by joining *alternate* corners by the diagonals of the faces we get two tetrahedra such that each edge of one meets two opposite edges of the other, and the figure possesses all the projective features of the most general pair of tetrahedra having this property.

Take an arbitrary tetrahedron of reference $XYZT$, and any point S whose homogeneous coordinates are x, y, z, t. Draw three lines through this point to meet the pairs of opposite edges, and on each line take the harmonic conjugate of S with respect to the intercept between the edges; in this way three new points P, Q, R are obtained, making in all the set of four

$$\begin{array}{ll} P, & (\ \ x, -y, -z, t\ ), \\ Q, & (-x,\ \ y, -z, t\ ), \\ R, & (-x, -y,\ \ z, t\ ), \\ S, & (\ \ x,\ \ y,\ \ z, t\ ). \end{array}$$

Then $PQRS$ and $XYZT$ are a pair of tetrahedra possessing the above property, for PS and QR meet both XT and YZ, and so on; they are the most general pair, for the preceding harmonic construction is deduced from the fact that, by hypothesis, any face of one tetrahedron cuts the other in a complete quadrilateral whose diagonals are the edges in that face. When one tetrahedron is given the other is determined by any one of its corners. Tetrahedra so related are said to be *desmic* and to belong to a *desmic system*.

These two tetrahedra possess further the remarkable property of being in *fourfold perspective*; for the lines PT, QZ, RY, SX are concurrent in $(-x, y, z, t)$, and so on. Thus there are four centres of perspective

$$\begin{array}{ll} P', & (-x, \quad y, \quad z, \quad t\ ), \\ Q', & (\ x, \ -y, \quad z, \quad t\ ), \\ R', & (\ x, \quad y, \ -z, \quad t\ ), \\ S', & (\ x, \quad y, \quad z, \ -t\ ), \end{array}$$

and we see that the points P', Q', R' are obtained from S' in the same way as P, Q, R were obtained from S, that is, by changing the signs of two coordinates; hence the tetrahedra $P'Q'R'S'$ and $XYZT$ belong to the same desmic system.

It has just been shown that the twelve corners of the three tetrahedra $PQRS$, $P'Q'R'S'$, $XYZT$ lie by threes on sixteen lines, and from this it follows that each pair of tetrahedra is in fourfold perspective. Now the property of being in fourfold perspective is an equally good definition of desmic tetrahedra and all the other properties can be deduced from this. We are thus led to include *three* tetrahedra in every desmic system.

None of the projective features of the figure are lost by taking X, Y, Z to be the infinite ends of a frame of rectangular axes

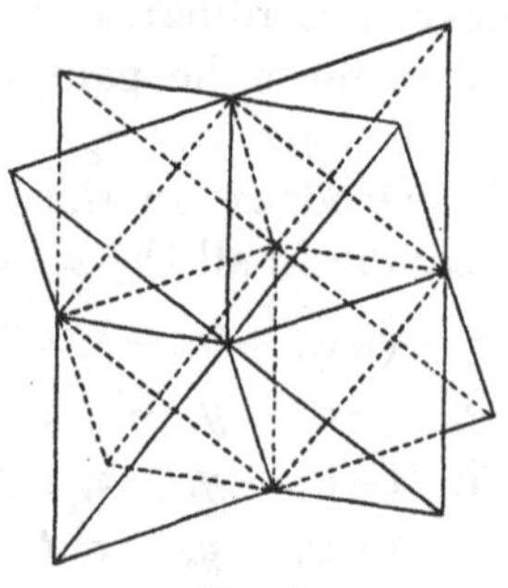

FIG. 1.

meeting in an origin T, and by supposing that $x=y=z$; then P, Q, R become the *images* of S in the axes, and P', Q', R', S' become the images of S in the planes of reference and in the origin respectively. We have, in fact, the corners of a cube, and the diagonals of the faces are the edges of the regular tetrahedra $PQRS$ and $P'Q'R'S'$. The figure is one that is easily conceived, and its desmic properties are readily discerned by geometrical intuition. Sets of parallel edges of the cube show the tetrahedra in three perspective aspects, and the diagonals show that the centre is also a centre of perspective. Of the intersections of the edges, six are the centres of the faces of the cube and are the corners of a regular octahedron. The remaining six are at infinity.

It is interesting to notice that these twelve points of intersection are the corners of another desmic system of tetrahedra formed with the same eighteen edges; in the figure the three new tetrahedra are infinite and wedge-shaped, each being formed by two opposite faces of the cube and the planes containing their parallel diagonals.

Since the figure is defined by intersecting lines, it is self-reciprocal, and to every point-theorem there is a corresponding plane-theorem; in particular, the faces of any two tetrahedra can be paired in four ways so that the lines of intersection lie in a face of the third tetrahedron. The geometrical properties of the figure are deducible from the identity

$$(a-b-c+d)(-a+b-c+d)(-a-b+c+d)(a+b+c+d)$$
$$+(-a+b+c+d)(a-b+c+d)(a+b-c+d)(a+b+c-d)$$
$$\equiv 16abcd,$$

in which the letters may be regarded as current coordinates of either a plane or a point.

The two following results may be taken as examples.

The twelve centres of similitude of four spheres are the corners of a desmic system.

If three tetrahedra belong by pairs to different desmic systems, the remaining tetrahedra of the three systems belong to another system.

Desmic systems were first investigated by Stephanos* and are so named because the three tetrahedra belonging to one system are members of a pencil (faisceau, δεσμός) of quartic surfaces. The general surface of the pencil has twelve nodes and is the subject of a memoir by Humbert†; its equation may be written in the symmetrical form

$$\lambda(x^2t^2+y^2z^2)+\mu(y^2t^2+z^2x^2)+\nu(z^2t^2+x^2y^2)=0,$$

where

$$\lambda+\mu+\nu=0,$$

and the three tetrahedra are

$$(x^2-t^2)(y^2-z^2)=0, \quad (y^2-t^2)(z^2-x^2)=0, \quad (z^2-t^2)(x^2-y^2)=0,$$

the sum of the left sides being identically zero. For further details and references see a paper by Schroeter‡; an application to Spherical Trigonometry is given by Study§.

* *Darboux Bulletin* (1879), sér. 2, III, 424.

† *Liouville* (1891), sér. 4, VII, 353.

‡ *Crelle* (1892), CIX, 341.

§ *Mathematical Papers*, Chicago Congress (1893), p. 382.

§ 2. THE GROUP OF REFLEXIONS.

A *group of operations* is a set of operations such that the resultant of any number taken in any order is an operation of the set*; in particular the repetition of a single operation any number of times is equivalent to some member of the group.

The fact that by successive reflexions in the axes only a finite number of points are obtained from one arbitrary point shows that the operations of reflecting belong to a group. Considered algebraically the operations consist in changing the signs of two of the coordinates. Let the symbol A denote the operation of changing the signs of y, z, and therefore of changing S into P. The repetition of the operation A changes P back into S again, and this is expressed by the symbolic equation

$$A^2 = 1.$$

Here 1 denotes the *identical operation*, which does not alter the position of the point to which it is applied; we infer that it must be included in every group to which A belongs. Similarly let the operation B change the signs of z, x, and C those of x, y. Then $B^2 = 1$ and $C^2 = 1$. Further, if B and C are performed successively in either order the result is the same as when y and z, but not x, are changed in sign; this is expressed by the symbolic equations

$$BC = CB = A.$$

In other words, B and C are *permutable* and their *product* is A. These equations, with others deduced by symmetry, are sufficient to show that the four operations

$$1,\ A,\ B,\ C$$

form a *group*, for any combination of them can be reduced to one of themselves. The *multiplication table*

	A	B	C
A	1	C	B
B	C	1	A
C	B	A	1

is a convenient way of representing the equations

$$A^2 = B^2 = C^2 = 1,$$

$$BC = CB = A,\quad CA = AC = B,\quad AB = BA = C.$$

* For a complete definition and fuller explanations see Burnside's *Theory of Groups*, Chap. II.

It may happen that some of the members of a group form a group by themselves. In this case the smaller group is called a *subgroup* of the larger. For example (1, A) is a subgroup of (1, A, B, C); (1, B) and (1, C) are also subgroups.

When two permutable groups are given, a third group can be obtained by combining the members of one with the members of the other in all possible ways, and is called the *product* of the first two groups. For example, the group of reflexions is the product of any two of its subgroups. The *order* of the product group, that is, the number of its members, is the product of the orders of the first two groups.

§ 3. THE 16_6 CONFIGURATION.

In space of three dimensions a point or a plane may be represented by four symbols α, β, γ, δ used homogeneously. The condition of *incidence* of two elements (α, β, γ, δ) and (α', β', γ', δ') of different kinds may be taken to be

$$\alpha\alpha' + \beta\beta' + \gamma\gamma' + \delta\delta' = 0.$$

On account of the perfect reciprocity between point and plane in projective geometry, every theorem that will be proved has its correlative theorem: it will not be necessary to state the second result in every case.

It is immediately verifiable that the plane

$$(\ \alpha,\quad \beta,\quad \gamma,\quad \delta\ )$$

contains the six points

$$\begin{array}{llll}
(\ \ \delta, & -\gamma, & \ \ \beta, & -\alpha\ ), \\
(\ \ \delta, & \ \ \gamma, & -\beta, & -\alpha\ ), \\
(\ \ \gamma, & \ \ \delta, & -\alpha, & -\beta\ ), \\
(-\gamma, & \ \ \delta, & \ \ \alpha, & -\beta\ ), \\
(-\beta, & \ \ \alpha, & \ \ \delta, & -\gamma\ ), \\
(\ \ \beta, & -\alpha, & \ \ \delta, & -\gamma\ ),
\end{array}$$

and this is the foundation of the configuration; the preceding six incidences are true for all values of the symbols, and we may therefore substitute the members of any of the last six rows for α, β, γ, δ respectively and so obtain other sets of incidences. It will be found that only sixteen different points and sixteen different planes can be obtained in this way, and this is due to the fact that the operations of permuting and sign-changing involved in these substitutions belong to a *group*. In order to explain the formation of this group we must introduce symbols for its members.

§ 4. THE GROUP OF SIXTEEN OPERATIONS.

Let A denote the operation of interchanging α, δ and at the same time β, γ, each letter carrying its sign with it; similarly B interchanges β with δ and γ with α, and C interchanges γ with δ and α with β. Further let A' denote the operation of changing the signs of β, γ, B' those of γ, α, and C' those of α, β. Then A', B', C' belong to the group of reflexions which has already been considered, and it is easily seen that A, B, C satisfy symbolic equations of precisely the same form; in other words $(1, A, B, C)$ and $(1, A', B', C')$ are two groups having similar multiplication tables. Since change of order is independent of change of sign, all the members of one group are permutable with those of the other, for example $AB' = B'A$, and consequently the groups themselves are said to be permutable.

By combining the members of these two permutable groups in all possible ways we obtain a set of sixteen operations which evidently form a *group* containing $(1, A, B, C)$ and $(1, A', B', C')$ as subgroups. All the sixteen planes of the configuration are obtained by operating on any one of them, say $(\alpha, \beta, \gamma, \delta)$, with the members of the group, and the six points lying in each plane are obtained by operating on the set given in § 3 with the corresponding member of the group; for, the condition of incidence is unaffected when the same operation is performed on point and plane. We may clearly use the symbol of operation to denote the point or plane obtained from $(\alpha, \beta, \gamma, \delta)$ by that operation, thus (1) denotes $(\alpha, \beta, \gamma, \delta)$ and (AB') denotes $(\delta, -\gamma, \beta, -\alpha)$, and so on; we have seen that the plane (1) or $(\alpha, \beta, \gamma, \delta)$ contains the points

$$(AB'),\ (AC'),\ (BC'),\ (BA'),\ (CA'),\ (CB'),$$

and we deduce that the plane (A), or $(\delta, \gamma, \beta, \alpha)$, contains the points

$$(B'),\ (C'),\ (CC'),\ (CA'),\ (BA'),\ (BB'),$$

and so on.

The group of sixteen operations, which will be referred to simply as *the group*, contains many subgroups. Any two operations and their product form, with the identical operation, a subgroup: two examples are $(1, AB', BC', CA')$ and $(1, AC', BA', CB')$. Further the group can be arranged in many ways as the product of two subgroups; one arrangement arises from the definition and another from the two preceding subgroups. These are shown by the multiplication tables

1	A'	B'	C'
A	AA'	AB'	AC'
B	BA'	BB'	BC'
C	CA'	CB'	CC'

1	AC'	BA'	CB'
AB'	A'	CC'	B
BC'	C	B'	AA'
CA'	BB'	A	C'

It is an easy exercise to verify the following table:

order of subgroup = 2, 4, 8,

number of subgroups = 15, 35, 15.

§ 5. THE INCIDENCE DIAGRAM.

We have thus found the coordinates of sixteen points and sixteen planes such that six points lie in each plane and six planes pass through each point. The most general 16_6 configuration, which is defined by these properties, can be reduced to the preceding form by a proper choice of coordinates.

The whole scheme can be exhibited very compactly by the following artifice. Since the subgroups $(1, A, B, C)$ and $(1, A', B', C')$ obey the same laws, they may be represented by the same symbols: the members of the two subgroups will be distinguished by the position they occupy in a compound symbol. Every member of the product group will be represented by a two-letter symbol in which the first letter will represent a member of $(1, A, B, C)$ and the second a member of $(1, A', B', C')$. The operations of each of these subgroups will be denoted by

$$d, a, b, c,$$

so that in either position d represents the identical operation. The multiplication table

	d	a	b	c
d	d	a	b	c
a	a	d	c	b
b	b	c	d	a
c	c	b	a	d

which is fundamental for this representation, applies to both the first and the second letters in a compound symbol, and the table showing the product of the two subgroups $(1, AB', BC', CA')$ and $(1, AC', BA', CB')$ takes the form

dd	ac	ba	cb
ab	da	cc	bd
bc	cd	db	aa
ca	bb	ad	dc

The sixteen symbols, which in the first instance denote operations, can, as is explained above, be used to denote both points and planes; it will be found that no confusion arises from this, but that, on the contrary, the duality of the configuration is clearly brought out by this nomenclature. Now any row is obtained from any other row by one of the operations ab, bc, ca, and any column from any other column by one of the operations ac, ba, cb. Since the plane (dd) contains the points represented by the other symbols in the same row and column, it follows that the *six elements incident with any element are given by the row and column containing that element.*

This property of the table is not lost if the rows are permuted in any manner, and also the columns. This, as well as the group property, shows that all the elements are of equal importance in the configuration, although the notation isolates (dd), or $(\alpha, \beta, \gamma, \delta)$. To bring this out more clearly the symbols in the table will frequently be replaced by dots, and then we shall have an *incidence diagram* which will be of great use for indicating

•	o	•	•		•	•	•	•
o	×	o	o		•	×	•	o
•	o	•	•		•	o	•	×
•	o	•	•		•	•	•	•

at a glance relations among the elements of the configuration. Thus, for example, in the first diagram the plane × contains the six points o, and the second diagram shows that any two planes have two points in common.

§ 6. LINEAR CONSTRUCTION FROM SIX ARBITRARY PLANES.

By means of the incidence diagram it is easy to prove that the 16_6 configuration can be linearly constructed from six arbitrary planes, and also, reciprocally, from six arbitrary points. It is convenient to use two diagrams, one for planes and the other for points; each is an incidence diagram for the elements contained in it, and two elements, one from each diagram, are incident if they lie on corresponding rows, or columns, but not both.

×	×	×	·		o	o	o	o
×	·	·	·		·	o	o	·
·	×	·	·		o	·	o	·
·	·	×	·		o	o	·	·

Let the first diagram represent any six planes; the positions of the crosses make no suppositions as to the linear dependence of the planes, for the diagram does not indicate that more than three planes pass through the same point. It is required to fill in the remaining ten places, if possible, so as to complete the incidence diagram and obtain a 16_6 configuration. The noughts in the second diagram represent ten of the twenty points of concurrence of the six planes, taken by threes: for example, the three crosses in the first row determine the last nought of the first row. Now every row and column in the second diagram, taken together, contain enough noughts to determine a plane of the configuration; in this way the remaining ten planes are found and the first diagram may be completed.

Hence a 16_6 configuration can be constructed from six arbitrary planes in at least one way, and therefore involves eighteen arbitrary constants. Now the system considered in § 3 contains the three ratios $\alpha : \beta : \gamma : \delta$ and fifteen constants implied in the choice of a particular set of homogeneous coordinates. We infer that the general configuration can be represented in this way.

We shall now investigate the preceding process of constructing the configuration in greater detail, and prove that six given planes determine *twelve* configurations.

Let five planes in general position be denoted by 1, 2, 3, 4, 5, their lines of intersection by two-figure symbols and their common points by three-figure symbols. There are twelve different cyclical arrangements of the planes and each gives a skew pentagon formed by the intersections of the planes taken in order. Thus corresponding to the arrangement

1 2 3 4 5 1

there is a pentagon with sides

12 23 34 45 51

and corners

123 234 345 451 512 ....*P*.

Each side contains, besides two corners of the pentagon, one other point of the system, where it meets the *opposite* plane, making the set

124 235 341 452 513 ....Q,

and these, when arranged in the order

135 352 524 241 413................P',

are the corners of the pentagon corresponding to the cyclical arrangement of planes 13524. The relation between the pentagons P and P' is mutual, and so the twelve pentagons can be divided into six pairs, the members of each pair being mutually inscribed and circumscribed.

We next prove that the pentagons whose corners are P and Q, taken in the order given, are so related that when they are projected from any point on to any plane, five intersections of pairs of sides are collinear. Giving the projections the same names as the points and lines in space, we see that the sets of points

341 123 513

and 512 452 235

are collinear, lying on 13 and 25 respectively. Therefore, by Pascal's theorem, the intersections of

(513, 512), or 51, and (235, 341)

(512, 123), or 12, and (341, 452)

(123, 235), or 23, and (452, 513)

are collinear, and similarly for the other pairs of sides. Hence the theorem is proved.

Two skew pentagons, which are so related that the five lines from any point to meet pairs of corresponding sides are coplanar, are said to be in *lineal position*; we have now proved that the twelve pentagons formed by the intersections of five planes taken in different orders can be arranged in six pairs such that if the corners of one pentagon are taken alternately a new pentagon is formed which is in lineal position with the other member of the pair. In this way we get twelve pairs of pentagons in lineal position.

Conversely, instead of projecting from an arbitrary point, take any sixth plane 6; its intersections with the lines 12 and (134, 245), 23 and (245, 351) determine two lines meeting in a point which must be collinear with the points where 6 cuts 34 and (351, 412), and so on. Thus we have a property of any six planes 1, 2, 3, 4, 5, 6, namely that the five planes

I, (612, 134, 245)
II, (623, 245, 351)
III, (634, 351, 412)
IV, (645, 412, 523)
V, (651, 523, 134)

have a common point lying in the plane 6. This is one of the results which were proved above from the diagrams; when the names of the planes and points are inserted, the preceding theorem is made evident from the diagrams

Planes				Points			
1	2	4	I	245	134	612	412
5	·	·	IV	·	523	645	·
·	3	·	III	351	·	634	·
II	V	6	·	651	623	·	·

Of the twenty points of concurrence of the six planes only ten are used in the second diagram and no two involve all six planes. Each plane occurs five times and the scheme is based on isolating one, 6, and arranging the rest in a certain cyclical order 12345. Any of the other planes may be isolated and the rest arranged in appropriate order so as to lead to the same configuration, thus

1, 35624
2, 41635
3, 52641
4, 13652
5, 24613

the rule being to substitute 6 in 12345 in place of the isolated figure and to interchange the two figures not adjacent to it in the cyclical order.

Hence when six planes are arbitrarily given, a 16_6 configuration is determined when one is isolated and the rest arranged in cyclical order. The twelve different orders of these five planes lead to twelve distinct configurations, and since it is immaterial which of the six planes is isolated in the first instance, only twelve different configurations can be obtained from them.

Six points can be chosen out of the sixteen so that no three lie in a plane of the configuration in many ways; the diagrams are of two types:

(1) four points form a rectangle and the others lie on different rows and columns,

(2) three points lie in a row and the others on different rows and the same columns.

Only one half of the configuration can be linearly constructed from a hexad of type (1).

Of the 720 ways of permuting the names of the points of type (2), sets of sixty lead to the same ten planes and only twelve differently named configurations can be obtained.

The fact that twelve 16_6 configurations can be linearly constructed from six given points was first proved by Weber*, so that the 192 different sets of six points in one configuration are called Weber hexads. The subject is treated synthetically by Reye† and Schroeter‡ who introduces pentagons in lineal position. Geiser§ examines in greater detail the groups of ten planes determined by six points and exhibits the configuration in an incidence diagram.

§ 7. SITUATION OF COPLANAR POINTS.

Let the diagram represent points, so that A, B, C, A', B', C' lie in one plane. $EFBC'$ and $EFB'C$ are seen to be plane quadrangles and therefore EF, BC', and $B'C$ meet in the common point P of these three planes. Similarly FD, CA', $C'A$ are concurrent in Q and DE, AB', $A'B$ in R. Hence the plane ABC

$$\begin{array}{cccc} A & B & C & \cdot \\ D & \cdot & \cdot & A' \\ \cdot & E & \cdot & B' \\ \cdot & \cdot & F & C' \end{array}$$

cuts the sides of the triangle DEF in points P, Q, R which are the intersections of opposite sides of the hexagon $AB'CA'BC'$. Since P, Q, R are collinear, it follows from Pascal's theorem that the hexagon is *inscribed in a conic.*

Since the six points in any one plane can be changed into the points in any other plane by a linear transformation belonging to the group, it follows that *the projective situation of the points on every conic is the same.*

* *Crelle* (1878), LXXXIV, 332.

† *Crelle* (1879), LXXXVI, 209.

‡ *Crelle* (1887), C, 231.

§ *Vierteljahrschrift der naturforschenden Gesellschaft in Zürich* (1896), Jahrgang 41, Teil II, p. 24.

The duality of the configuration shows that the six planes through any point touch a quadric cone and the situation of the lines of contact is projectively equivalent to that of the six points on a conic.

Further, the twisted cubic curve determined by the six points $ABCDEF$ is projected from D into a conic in the plane ABC passing through $ABCQR$. The pencil $A[BCQR]$ is the same as $A[BCC'B']$, showing that the ranges $BCC'B'$ on the conic and $BCFE$ on the cubic have the same cross ratio. Similarly the other ranges of four points may be compared.

From their position in the diagram we see that $ABCDEF$ are six points from which the whole configuration can be linearly constructed; and it has just been proved that their situation on the twisted cubic through them is projectively equivalent to that of the points $ABCA'B'C'$ on the conic through them. The same is true of all the Weber hexads and all the conics, and three independent cross ratios of the six parameters which give the positions of the six points on any conic may be taken to be the absolute invariants of the configuration, being unaltered by any linear transformation of coordinates.

CHAPTER II.

THE QUARTIC SURFACE.

§ 8. THE QUARTIC SURFACE WITH SIXTEEN NODES.

Every surface can be regarded either as a locus of points or as an envelope of planes. It is convenient to give preponderance to the former view, so that the *order* of a surface is one of its most distinguishing features. Carrying this idea further, it is natural to classify surfaces of given order by the *point singularities* which they may possess. A singular point is one in the neighbourhood of which the surface ceases to be approximately flat; there may be a locus of such points, giving a *singular line*, or the singular points may be isolated. Among surfaces of the second order the former case is illustrated by a pair of planes, and the latter by a cone.

The surface usually known by the name of its first investigator, Kummer*, belongs to those surfaces of the fourth order which have isolated singularities. The only kind of point singularity which will in general be considered is that of a *node*, characterised geometrically by the fact that the tangent lines at it generate a quadric cone instead of a plane, and algebraically by the absence of terms of the first degree from the equation in point coordinates when the node is taken as origin.

The reciprocal singularity, which will occur with equal frequency, is that of a *trope*, or singular tangent plane. It is cut by consecutive tangent planes in lines which envelope a conic instead of forming, as usual, a plane pencil at the point of contact. The plane therefore touches the surface all along a conic instead of at a single point. The conic of contact of a trope will sometimes be called a *singular conic* of the surface.

The number of nodes which a surface of order n can have is limited by its *class*. The points of contact of tangent planes

* Ernst Eduard Kummer, Professor of Mathematics at Berlin, 1856; born 1810, died 1893. See *Berliner Monatsberichte*, (1864), pp. 246, 495; (1865), p. 288. *Berlin. Akad. Abhandl.* (1866), p. 1.

through any two points A and B lie on the polar surfaces of A and B, which are of order $n-1$, and hence the number of them is $n(n-1)^2$. But by considering a penultimate form of surface it appears that two of these points coincide with every node; hence the formula for the class is $n(n-1)^2 - 2\delta$, where δ is the number of nodes. Putting $n = 4$ we get $36 - 2\delta$ for the class, showing that δ cannot be greater than 16, for if the class were 2 the surface would be a quadric.

That a quartic surface can actually have as many as sixteen nodes will be proved in § 10 by constructing its equation. Assuming this for the present, we proceed to deduce some of the properties of this configuration of nodes from the elementary geometry of the surface.

At any node there is a quadric cone of tangent lines which is touched by the enveloping cone *from* the node along six generators, namely those tangent lines at the node which have four-point contact with the surface, and are found by equating to zero the terms of degrees 2 and 3 in the equation referred to the node as origin. The enveloping cone is of order 6. It has nodal lines passing through the fifteen other nodes, for a node on a surface is a node on its apparent contour. But if a sextic cone has fifteen nodal lines, it must break up into six planes. Hence the enveloping cone from any node consists of six planes touching the quadric tangent cone at the node and containing the remaining fifteen nodes on their lines of intersection. Since any one plane is cut by five others, six nodes lie on each plane. Consider the section of the surface by one of these planes; every line drawn in this plane through the node is a tangent line and meets the surface in one point distinct from the node, namely its point of contact. Hence the section must be a conic passing through six nodes, that is, the plane touches the surface all along a conic, and is therefore a trope. The complete section of the surface by a trope is a conic counted twice; since this passes through six nodes, the trope must touch the six quadric tangent cones along generators which are tangents to the singular conic.

We thus see that if a quartic surface has sixteen nodes, they are situated by sixes in tropes each of which touches the surface along a conic. Since six tropes pass through each node their number is also 16. The nodes and tropes form a 16_6 configuration like that considered in § 3 and is of the most general character. When the singularities are given the surface is unique, for sixteen conics on it are given.

§ 9. NOMENCLATURE FOR THE NODES AND TROPES.

Although the names given to the points and planes in § 5 are suitable for a symmetrical treatment of the configuration and for exhibiting the interchanges that take place under the operations of the group, yet it is often desirable to isolate a particular element, or a particular set of six elements. This is done in the following way.

One node is called 0 and the six tropes through it are 1, 2, 3, 4, 5, 6. The two-figure symbols 12, 13 ... 56 denote the fifteen nodes other than 0 lying on the intersections of these planes. We have to find how these nodes are arranged on the remaining ten tropes.

When two triangles circumscribe a conic, their six corners lie on a conic, and a corresponding theorem is true for a cone. Since the six tropes through 0 touch the tangent quadric cone to the surface at 0, it follows that the six nodes

13, 35, 51, 24, 46, 62

lie on a quadric cone whose vertex is 0. By partitioning the six figures 1, 2, 3, 4, 5, 6 into two sets of three in all possible ways we get ten different cones, and in general, no other cones with vertices at 0 contain sets of six nodes. Now the six nodes in any one of ten tropes are projected from 0 by a quadric cone. Accordingly, the preceding six lie on a trope which may be called (135 . 246), or simply 135, or 246.

Thus we have two nomenclatures for the sixteen elements of a 16_6 configuration; in the first a single element is isolated and the remaining fifteen are named after the combinations of six figures taken two at a time: in the second, six elements—either concurrent tropes or coplanar nodes—are named by single figures and the remaining ten by the partitions of these six into two sets of three. The two nomenclatures may be used simultaneously, one for the nodes, and the other for the tropes of the same configuration, as above, and then we have the following types of incidences:

0	is incident with	1,	2,	3,	4,	5,	6,
12	„ „ „	2,	1,	123,	124,	125,	126,
1	„ „ „	0,	12,	13,	14,	15,	16,
123	„ „ „	23,	31,	12,	56,	64,	45.

All these incidences are clearly indicated in a pair of square diagrams representing points and planes respectively. An element of one diagram is incident with the elements of the other diagram lying in the same row and the same column, but not both. The pair of diagrams can be constructed in ten essentially different ways, corresponding to the different partitions of six figures into two threes, one of which is the following,

$\frac{135}{246}$	2	4	6
1	$\frac{146}{235}$	$\frac{162}{435}$	$\frac{124}{635}$
3	$\frac{346}{251}$	$\frac{362}{451}$	$\frac{324}{651}$
5	$\frac{546}{213}$	$\frac{562}{413}$	$\frac{524}{613}$

0	46	62	24
35	12	14	16
51	32	34	36
13	52	54	56

§ 10. THE EQUATION OF THE SURFACE.

With homogeneous coordinates x, y, z, t let the node 0 be (0, 0, 0, 1), then the equation must have the form

$$\phi_2 t^2 + 2\phi_3 t + \phi_4 = 0,$$

where ϕ_s is homogeneous and of degree s in x, y, z.

The quadric tangent cone at 0 is

$$\phi_2 = 0,$$

and the enveloping sextic cone from 0 is

$$\phi_3^2 - \phi_2 \phi_4 = 0,$$

which must break up into six planes. Both of these equations are known when the nodes are given. Further, if one of the tropes not passing through 0 is taken to be $t = 0$, the equation

$$\phi_4 = 0$$

represents the repeated conic passing through the nodes in that trope, and hence ϕ_4 is known when the nodes are given, except as to a numerical factor.

Choose the coordinates so that

$$\phi_2 \equiv y^2 - xz.$$

It is convenient to represent any generator of the cone $\phi_2 = 0$ in terms of a parameter u by the equations

$$x = y/u = z/u^2.$$

Let k_1, k_2, k_3, k_4, k_5, k_6 be the values of u giving the lines of contact of the six tropes concurrent in 0; then their equations are

$$x_s \equiv k_s^2 x - 2k_s y + z = 0, \qquad (s = 1, 2, 3, 4, 5, 6)$$

and we must have the identity

$$\phi_3^2 - \phi_2\phi_4 = \lambda x_1 x_2 x_3 x_4 x_5 x_6,$$

where λ is an undetermined constant.

We shall take $t=0$ to be the plane (123.456) and then $\phi_4 \equiv \psi_2^2$ where $\psi_2 = 0$ is the conic in $t=0$ passing through the nodes 23, 31, 12, 56, 64, 45. It is convenient to represent any point in the plane $t=0$ by the parameters u, v of the two tangents from it to the conic $y^2 = xz$. Thus we write

$$x = 2y/(u+v) = z/uv,$$

then $$k_s^2 x - 2k_s y + z = x(u-k_s)(v-k_s),$$

and the node 12 is given by $u=k_1$, $v=k_2$ or by $u=k_2$, $v=k_1$.

Consider now the equation

$$(u-k_1)(u-k_2)(u-k_3)(v-k_4)(v-k_5)(v-k_6)$$
$$-(v-k_1)(v-k_2)(v-k_3)(u-k_4)(u-k_5)(u-k_6)=0.$$

After division by $u-v$ it becomes symmetric in u and v, the highest term being u^2v^2, and is therefore the equation of a conic expressed in terms of the new coordinates u, v. It is obviously satisfied by the six points

$$(u,\ v) = (k_2,\ k_3),\ (k_3,\ k_1),\ (k_1,\ k_2),\ (k_5,\ k_6),\ (k_6,\ k_4),\ (k_4,\ k_5),$$

and hence is the conic $\psi_2 = 0$. Introduce the abbreviations

$$u_s = u - k_s, \quad v_s = v - k_s,$$

then $$\psi_2 = 2\mu x^2 (u_1u_2u_3v_4v_5v_6 - v_1v_2v_3u_4u_5u_6)/(u-v),$$

where μ is an undetermined constant; further

$$4\phi_2 = x^2(u-v)^2,$$

and $$\phi_3^2 - \phi_2\phi_4 = \lambda x^6 u_1u_2u_3u_4u_5u_6v_1v_2v_3v_4v_5v_6,$$

so that

$$x^{-6}\phi_3^2 = \mu^2(u_1u_2u_3v_4v_5v_6 - v_1v_2v_3u_4u_5u_6)^2 + \lambda u_1u_2u_3u_4u_5u_6v_1v_2v_3v_4v_5v_6,$$

and since this expression must be the square of a symmetric function of u and v we must have $\lambda = 4\mu^2$, which gives

$$x^{-3}\phi_3 = \pm\mu(u_1u_2u_3v_4v_5v_6 + v_1v_2v_3u_4u_5u_6).$$

On substituting these values of ϕ_2, ϕ_3, $\phi_4 = \psi_2^2$ in the equation of the surface we find

$$\phi_2 t = -\phi_3 \pm \sqrt{\phi_3^2 - \phi_2\phi_4},$$

$$(u-v)^2 t/x = \mp\mu(u_1u_2u_3v_4v_5v_6 + v_1v_2v_3u_4u_5u_6)$$
$$\pm 2\mu\sqrt{u_1u_2u_3u_4u_5u_6v_1v_2v_3v_4v_5v_6}$$
$$= \mp\mu(\sqrt{u_1u_2u_3v_4v_5v_6} \mp \sqrt{v_1v_2v_3u_4u_5u_6})^2.$$

In this parametric representation of the surface with given nodes there is apparently an arbitrary constant μ, but it will be noticed that the only data that have been used are one node 0 and the six tropes through it. The actual position of any other node determines the value of μ. There is no loss of generality in taking $\mu = 1$, and determining the signs of the radicals so that the coordinates of any point on the general sixteen-nodal quartic can be expressed in terms of two parameters u, v by the equations*

$$2y/x = u + v,$$
$$z/x = uv,$$
$$t/x = (u - v)^{-2}(\sqrt{u_1 u_2 u_3 v_4 v_5 v_6} + \sqrt{v_1 v_2 v_3 u_4 u_5 u_6})^2,$$

and the last expression is one of ten similar forms corresponding to the ten tropes which do not contain the node $x = y = z = 0$.

If the equations of the sixteen tropes are

$$x_1 = 0, \text{ etc.}; \quad x_{123} \equiv x_{456} = 0, \text{ etc.},$$

we may write, omitting a factor of proportionality,

$$x_s = u_s v_s, \qquad (s = 1, 2, 3, 4, 5, 6)$$
$$(u - v)\sqrt{x_{123}} = \sqrt{u_1 u_2 u_3 v_4 v_5 v_6} + \sqrt{v_1 v_2 v_3 u_4 u_5 u_6},$$

and so on. Then it is easy to verify that

$$(k_2 - k_3)\sqrt{x_1 x_{234}} + (k_3 - k_1)\sqrt{x_2 x_{314}} + (k_1 - k_2)\sqrt{x_3 x_{124}} = 0,$$

and

$$(k_1 - k_4)(k_2 - k_3)\sqrt{x_{235} x_{236}} + (k_2 - k_4)(k_3 - k_1)\sqrt{x_{315} x_{316}} + (k_3 - k_4)(k_1 - k_2)\sqrt{x_{125} x_{126}} = 0,$$

which are two of the many irrational forms in which the equation of the surface can be expressed.

§ 11. THE SHAPE OF THE SURFACE.

It has been shown that the six tropes through any node cut any other trope in two triangles whose corners are nodes. Hence if the six nodes in any plane are partitioned into three pairs, there is another trope through each pair, and these three tropes meet in a point which is *not* a node. The four planes are the faces of a tetrahedron† and contain singular conics intersecting by pairs on its edges. The quadric which can be drawn through two of these conics and one of the remaining two nodes cuts each

* Darboux, *Comptes Rendus* (1881), XCII, 1493.

† Named after Göpel.

of the other two conics in five points and therefore contains them entirely. Thus the four singular conics lie on a quadric.

Let ordinary tetrahedral coordinates be used and let the quadric be $\phi = 0$, then the quartic surface referred to this tetrahedron of tropes must be

$$xyzt = k\phi^2,$$

where the value of k depends on the coefficients in ϕ. When $k = 0$ we have the four faces of the tetrahedron, and when k is small, which is the case most easily realised, the surface lies near the faces. There is a distinction, important from a metrical point of view, between the cases when k is positive and negative. In the former case the product $xyzt$ is positive and the point (x, y, z, t) lies either *within* the tetrahedron or in one of the wedge-shaped regions opposite the edges. Fig. 2 shows the region for which

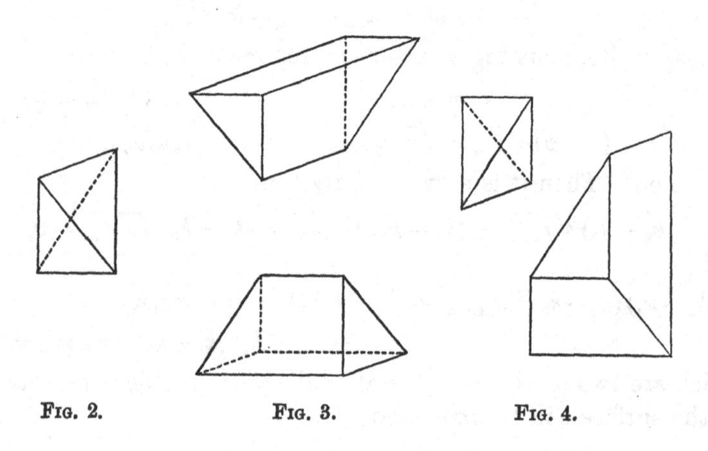

FIG. 2. FIG. 3. FIG. 4.

x, y, z, t are all positive, and fig. 3 shows the region obtained from it by the substitution

$$x'/x = y'/y = -z'/z = -t'/t,$$

and indicates that the two wedge-shaped pieces should be taken together as being continuous at infinity and forming one "tetrahedron." Again fig. 4 shows the effect of the substitution

$$-x'/x = y'/y = z'/z = t'/t,$$

and gives a "tetrahedron" of which one corner is separated from the other three by the plane at infinity. Hence when k is negative the sheets of Kummer's surface lie in the regions opposite the corners and the faces.

In order to realise conveniently the shape of the surface we suppose the tetrahedron to be regular and the quadric $\phi = 0$ to be a concentric sphere*. The sphere cuts the edges in twelve nodes lying by sixes on four circles in the faces. They lie by fours on twelve other planes which intersect by sixes in the corners of a similarly situated regular tetrahedron. We thus have sixteen nodes and sixteen tropes. First let the sphere be smaller than the circumsphere of the tetrahedron, then the second tetrahedron is situated inside the first and we have the case $k > 0$.

The four circles on the surface of the sphere $\phi = 0$ determine four triangles joined by six lunes. Fig. 5 is an orthogonal pro-

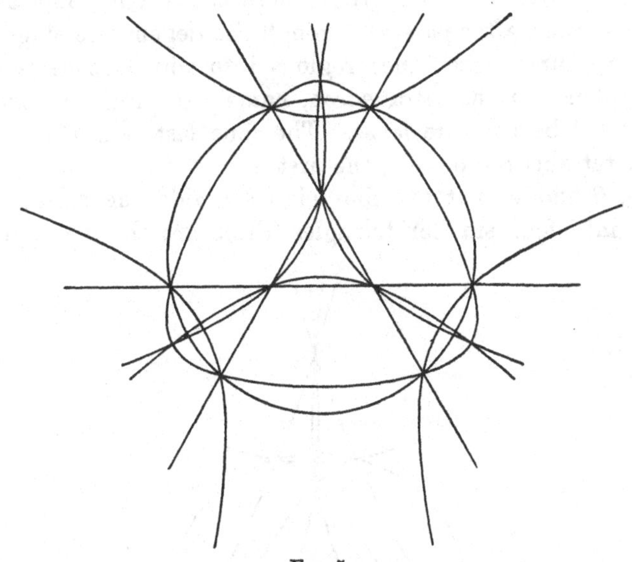

FIG. 5.

jection and shows one of these circles completely and parts of the other three. The twelve remaining conics are hyperbolas and six of these are drawn, of which three appear as straight lines. Only three nodes of the inner tetrahedron are shown.

Two nodes which are *adjacent* on one conic are adjacent on one other and belong to the same *piece* of the surface. The frontis-

* If $\phi \equiv x^2 + y^2 + z^2 + t^2 + \mu(xt + yt + zt + yz + zx + xy)$, k may be found from the conditions for a node, and the quartic surface is

$$\phi^2 + 4(\mu + 1)(\mu - 2)^2 xyzt = 0.$$

The radius r of $\phi = 0$, the circum-radius R, and μ are connected by

$$3(r^2 - R^2)\mu = 2(3r^2 + R^2).$$

piece, of which fig. 5 is a partial skeleton*, shows that the surface consists of a central four-cornered or *tetrahedral* piece attached by four other tetrahedral pieces, meeting the sphere $\phi=0$ in triangles, to six infinite wedge-shaped pieces, each of which contains two nodes and meets the sphere in a lune; those pieces which contain different branches of the same hyperbola may, by extending the notion of continuity, be paired so as to form three tetrahedral pieces each of which, like the central piece, has one node in common with each of the other four pieces.

As the sphere increases and approaches the circumsphere the surface approaches the four faces of the tetrahedron, the various portions of it remaining within the tetrahedron or in the regions beyond the edges. When the sphere still further increases, the quartic surface, after passing through the degenerate stage of four planes, appears in the other regions into which space is divided by the planes of the tetrahedron, namely the regions beyond the corners and beyond the faces. The four last nodes now form a regular tetrahedron outside the first.

Fig. 6 shows that the four circles divide the surface of the sphere into four smaller triangles (three are shown), four larger

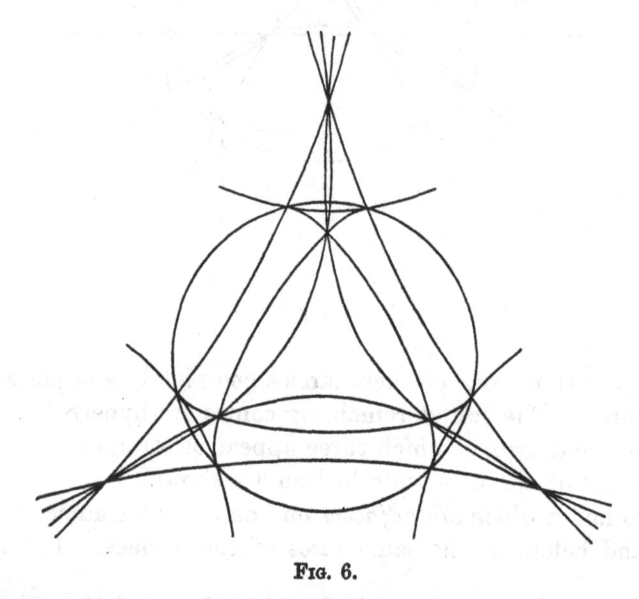

FIG. 6.

* Models of the sixteen conics passing by sixes through sixteen points are easily made out of coiled steel wire, beginning with the four circles. They have the advantage over plaster models that the surface is *transparent* and does not hide alternate arcs of conics.

triangles (one is shown), and six quadrangles. On each of the smaller triangles stands a tetrahedral piece attached at one node to an infinite conical piece; on each of the larger triangles stands an infinite piece containing three nodes which must be regarded as continuous at infinity with the opposite conical piece.

These two forms of Kummer's surface are equivalent from a descriptive point of view, and differ only in their relation to the plane at infinity. In both cases there are two sets of four tetrahedral pieces, each piece of one set being attached by a node to each piece of the other set.

CHAPTER III.

THE ORTHOGONAL MATRIX OF LINEAR FORMS.

§ 12. PRELIMINARY ACCOUNT OF MATRICES.

A *matrix** is simply a set of independent elements arranged in a rectangular array. In the abstract theory it is usual to indicate the row and column to which each element belongs by suffixes: thus we denote the element lying in the rth row and the sth column of a matrix a by a_{rs} and write

$$a = (a_{rs}).$$

In what follows we shall be concerned solely with *square* matrices, namely matrices which have the same number of rows as columns. This number is the *order* of the matrix.

Associated with every matrix a is its *conjugate* matrix $\bar{a}$, obtained by interchanging rows and columns: thus

$$\bar{a} = (\bar{a}_{rs}) \text{ implies } \bar{a}_{rs} = a_{sr}.$$

The addition of matrices of the same order is effected by adding corresponding elements: thus

$$a + b = c \text{ implies } a_{rs} + b_{rs} = c_{rs}.$$

The *product* of two matrices a and b of the same order n is a third matrix c defined by

$$c_{rt} = \sum_{s=1}^{n} a_{rs} b_{st} \qquad (r, t = 1, 2, \ldots n),$$

so that c has the same order as a and b. These n^2 equations are all written at once in the form

$$c = ab.$$

* For a completer account see Baker's *Abelian Functions*, p. 666, where numerous references are given.

The rule for forming the product of two matrices resembles the rule for multiplying two determinants, but it is important to notice that the *rows* of the first factor are taken with the *columns* of the second factor. From the definition it follows that the multiplication of matrices is associative, for if $ab = c$ and $bd = e$,

$$(cd)_{pq} = \Sigma_s c_{ps} d_{sq} = \Sigma_s \Sigma_r a_{pr} b_{rs} d_{sq} = \Sigma_r a_{pr} \Sigma_s b_{rs} d_{sq} = \Sigma_r a_{pr} e_{rq} = (ae)_{pq},$$

so that
$$cd = abd = ae;$$
but ab and ba are in general different matrices and the multiplication is not commutative.

The unit matrix E is defined by the equation

$$Ea = a$$

in which a is arbitrary; accordingly all the elements of E are zeros except those in the leading diagonal, which are unities. It then follows that $aE = a$ and $EE = E$. For these reasons E may often be replaced by 1, or omitted.

Let A_{rs} be the minor or cofactor of a_{rs} in $| a |$ the *determinant* of a, then

$$\Sigma_r A_{rs} a_{rs} = | a | = \Sigma_s A_{rs} a_{rs}, \text{ and } \Sigma_r A_{rs} a_{rt} = 0 = \Sigma_r A_{sr} a_{tr}, \text{ if } s \neq t.$$

Hence $A_{rs} / | a |$ is the srth element of a matrix which, when multiplied into a, gives E. It is called the *inverse* matrix and is written a^{-1}, so that we have

$$a^{-1} a = E = aa^{-1}.$$

When $| a | \neq 0$, either of these equations may be taken as the definition of the inverse matrix, for the n elements of any row of the first factor are given by n linear equations for which the determinant of coefficients is $| a |$. A matrix whose determinant vanishes has no inverse.

The following results are consequences of the definitions and may easily be verified by means of them. If

$$c = ab,$$

then the conjugate matrix $\bar{c} = \bar{b}\bar{a}$,

the inverse matrix $c^{-1} = b^{-1} a^{-1}$,

and the determinant $| c | = | a | \, | b |$.

A *row letter* x used in connection with matrices of order n is a symbol for n elements $(x_1, \ldots, x_n)$. Then ax is interpreted to mean n elements of which the first is $\Sigma a_{1s} x_s$, and xa means n elements of which the first is $\Sigma x_s a_{s1}$; hence x may be regarded in ax as a rectangular matrix of one column, and in xa as a matrix of one row. Accordingly, if y is another row or column,

$$xay = \Sigma\Sigma\, x_r a_{rs} y_s = \Sigma\Sigma\, y_s \bar{a}_{sr} x_r = y\bar{a}x.$$

The notation of matrices is in the first instance a device whereby suffixes and signs of summation may be omitted; in order to interpret a product of several matrices by supplying the suffixes, adjacent suffixes belonging to different letters must be the same, and the summations are with respect to them. But, further, the laws of operation are so framed that the system becomes a calculus, and we are able to manipulate matrices almost like single algebraical quantities. The chief use of matrices in this book is to obtain easily and to express clearly and shortly a great number of algebraic identities.

§ 13. ORTHOGONAL MATRICES.

An *orthogonal* matrix is defined as one which is the inverse of its conjugate; thus a is orthogonal if

$$a\bar{a} = E.$$

This condition is satisfied by the matrix of order 3 whose elements are the direction cosines of three mutually perpendicular lines, whence the name. The condition, when worked out, implies that the sum of the squares of the elements in any row is 1 and the sum of the products of corresponding elements in any two rows is 0. Since

$$\bar{a}a = a^{-1}a\bar{a}a = a^{-1}Ea = a^{-1}a = E,$$

it follows that the conjugate matrix is also orthogonal, and that the same conditions hold for the columns as for the rows. Thus from the equations

$$\Sigma_s\, a_{ps}{}^2 = 1, \quad \Sigma_s\, a_{ps} a_{qs} = 0$$

we have deduced

$$\Sigma_s\, a_{sp}{}^2 = 1, \quad \Sigma_s\, a_{sp} a_{sq} = 0.$$

Further, if the rows are permuted in any manner, and also the columns, the new matrix is orthogonal.

If a and b are both orthogonal and c is their product we have

$$a\bar{a} = E, \quad b\bar{b} = E, \quad c = ab, \quad \bar{c} = \bar{b}\bar{a},$$

whence it follows that

$$c\bar{c} = ab\bar{b}\bar{a} = aE\bar{a} = a\bar{a} = E.$$

Therefore *the product of two orthogonal matrices is an orthogonal matrix.* Further, since $a\bar{a} = E$, therefore $|a|\,|\bar{a}| = |E| = 1$; but $|a| = |\bar{a}|$, therefore $|a| = \pm 1$. Hence *the determinant of an orthogonal matrix is* ± 1. The theorem that *complementary two-rowed subdeterminants of a four-rowed orthogonal matrix are numerically equal* will be useful, and can easily be extended and generalised. If $a = (a_{rs})$, $\bar{a} = (\bar{a}_{rs}) = a^{-1}$, $\bar{a}_{rs} = a_{sr}$, then

$$\begin{bmatrix} \bar{a}_{11} & \bar{a}_{12} & \bar{a}_{13} & \bar{a}_{14} \\ \bar{a}_{21} & \bar{a}_{22} & \bar{a}_{23} & \bar{a}_{24} \\ 0 & 0 & 1 & 0 \\ 0 & 0 & 0 & 1 \end{bmatrix} \begin{bmatrix} a_{11} & a_{12} & a_{13} & a_{14} \\ a_{21} & a_{22} & a_{23} & a_{24} \\ a_{31} & a_{32} & a_{33} & a_{34} \\ a_{41} & a_{42} & a_{43} & a_{44} \end{bmatrix} = \begin{bmatrix} 1 & 0 & 0 & 0 \\ 0 & 1 & 0 & 0 \\ a_{31} & a_{32} & a_{33} & a_{34} \\ a_{41} & a_{42} & a_{43} & a_{44} \end{bmatrix}$$

and on equating the determinants of both sides

$$\begin{vmatrix} \bar{a}_{11} & \bar{a}_{12} \\ \bar{a}_{21} & \bar{a}_{22} \end{vmatrix} |a| = \begin{vmatrix} a_{33} & a_{34} \\ a_{43} & a_{44} \end{vmatrix}$$

or

$$a_{11}a_{22} - a_{12}a_{21} = \pm(a_{33}a_{44} - a_{34}a_{43}).$$

Since by permuting the rows and columns any pair of complementary subdeterminants can be brought into the position of any other pair, the theorem is proved.

§ 14. CONNECTION BETWEEN MATRICES AND QUATERNIONS.

The theory of four-rowed orthogonal matrices is intimately connected with *quaternions.* Introducing the complex units i, j, k defined solely by the equations

$$i^2 = j^2 = k^2 = -1,$$

$$jk = i = -kj, \quad ki = j = -ik, \quad ij = k = -ji,$$

we have associated with every quaternion

$$q = ix + jy + kz + t$$

a *conjugate* quaternion

$$q' = -ix - jy - kz + t,$$

obtained by reversing the *vector* part. Then the product qq' is *scalar,*

$$qq' = x^2 + y^2 + z^2 + t^2.$$

The multiplication of quaternions is associative but not commutative. It is easy to prove from the definitions that the conjugate of a product is the product of the conjugates taken in the reverse order: in symbols, if

$$r = pq$$

then the conjugate $\qquad r' = q'p'.$

It follows that $rr' = pqq'p' = p(qq')p' = (pp')(qq')$, because qq' is scalar.

This result when worked out in detail leads to an interesting theorem in matrices. Let

$$\begin{aligned} p &= i\alpha + j\beta + k\gamma + \delta, \\ q &= ix + jy + kz + t, \\ r = pq &= i(\ \delta, -\gamma,\ \beta, \alpha \between x, y, z, t) \\ &\quad + j(\ \gamma,\ \delta, -\alpha, \beta \between x, y, z, t) \\ &\quad + k(-\beta,\ \alpha,\ \delta, \gamma \between x, y, z, t) \\ &\quad + \ (-\alpha, -\beta, -\gamma, \delta \between x, y, z, t). \end{aligned}$$

Now $rr' = pp' . qq'$; hence the sum of the squares of these four linear forms is $(\alpha^2 + \beta^2 + \gamma^2 + \delta^2)(x^2 + y^2 + z^2 + t^2)$. Since this is true for all values of x, y, z, t, it follows that if $\alpha^2 + \beta^2 + \gamma^2 + \delta^2 = 1$, the matrix

$$\begin{bmatrix} \delta & -\gamma & \beta & \alpha \\ \gamma & \delta & -\alpha & \beta \\ -\beta & \alpha & \delta & \gamma \\ -\alpha & -\beta & -\gamma & \delta \end{bmatrix}$$

is orthogonal. Again, on arranging r differently,

$$\begin{aligned} r &= i(\alpha, \beta, \gamma, \delta \between\ t,\ z, -y,\ x) \\ &\quad + j(\alpha, \beta, \gamma, \delta \between -z,\ t,\ x,\ y) \\ &\quad + k(\alpha, \beta, \gamma, \delta \between\ y, -x,\ t,\ z) \\ &\quad + \ (\alpha, \beta, \gamma, \delta \between -x, -y, -z,\ t), \end{aligned}$$

whence, by similar reasoning, the matrix

$$\begin{bmatrix} t & z & -y & x \\ -z & t & x & y \\ y & -x & t & z \\ -x & -y & -z & t \end{bmatrix}$$

is orthogonal if $x^2 + y^2 + z^2 + t^2 = 1$.

§ 15. THE SIXTEEN LINEAR FORMS.

Instead of using a set of four symbols $(\alpha, \beta, \gamma, \delta)$ to denote a point or plane, we now supply current coordinates (x, y, z, t) and use a *linear form*, which, when equated to zero, gives the *equation* of the point or plane. Thus we write

$$\begin{aligned} (dd) &= (\alpha, \beta, \gamma, \delta \between\ x,\ \ y, z, t), \\ (bc) &= (\gamma, \delta, \alpha, \beta \between -x, -y, z, t), \end{aligned}$$

and so on, obtaining sixteen forms which are linear and homogeneous in each set of symbols $(\alpha, \beta, \gamma, \delta)$ and (x, y, z, t). We recall that the first letter of a two-letter symbol such as bc denotes a permutation and the second denotes a change of sign; in deducing the form (bc) from (dd) the first operation is performed upon $(\alpha, \beta, \gamma, \delta)$ and the second upon (x, y, z, t); with this convention the coefficient of t in every form has a positive sign. The same result is obtained when the compound operation is performed upon (x, y, z, t), thus

$$(bc) = (\alpha, \beta, \gamma, \delta \between z, t, -x, -y).$$

These forms are connected by a remarkable set of algebraic identities, and the geometrical interpretations of them lead to important theorems concerning the configuration. Before proceeding to develop these identities it is convenient to give the whole set of forms; they may be written down as the elements of a matrix which is the product of a matrix of coefficients and a matrix of coordinates. Thus from the definitions we have

$$\begin{bmatrix} (aa) & (ab) & (ac) & (ad) \\ (ba) & (bb) & (bc) & (bd) \\ (ca) & (cb) & (cc) & (cd) \\ (da) & (db) & (dc) & (dd) \end{bmatrix} = \begin{bmatrix} \delta & \gamma & \beta & \alpha \\ \gamma & \delta & \alpha & \beta \\ \beta & \alpha & \delta & \gamma \\ \alpha & \beta & \gamma & \delta \end{bmatrix} \begin{bmatrix} x & -x & -x & x \\ -y & y & -y & y \\ -z & -z & z & z \\ t & t & t & t \end{bmatrix}$$

where it may be noticed that the *rows* of the first factor are obtained from $(\alpha, \beta, \gamma, \delta)$ by the permutations a, b, c, d, and the *columns* of the second factor are obtained from (x, y, z, t) by the changes of sign denoted by a, b, c, d.

The sixteen forms are obtained from one of them (dd) by the group of sixteen operations which may now be regarded as linear transformations of the coordinates x, y, z, t. It is important to know the effect of these transformations on any other form; it is easy to see that any two forms are interchanged, except possibly as to sign, by the transformation whose symbol is the product of the symbols of the forms. If the product of the operations pq and $p'q'$ is $p''q''$ then the form (pq) is changed by the transformation $p'q'$ into $\pm(p''q'')$, and the lower sign must be taken if $(p'q)$ is one of the six forms

$$(ab), (ac), (bc), (ba), (ca), (cb),$$

in which the coefficient of δ has a minus sign.

The *linear* identities which exist among the sixteen forms have already been indicated, for they correspond to sets of four coplanar points. We infer that a set of six forms can be selected in sixteen different ways such that any four of the set are linearly connected.

It is sufficient to find the coefficients in the relations connecting one set of six forms, for then all the other relations can be deduced by operating on (x, y, z, t) with the transformations of the group.

§ 16. QUADRATIC RELATIONS.

Another important series of relations is obtained when we have shown that the sixteen linear forms, with proper signs, can be arranged as the elements of an *orthogonal matrix*. After making slight changes in the results of § 14, and under the hypotheses that $\alpha^2 + \beta^2 + \gamma^2 + \delta^2 = 1$, $x^2 + y^2 + z^2 + t^2 = 1$, which are of no importance since the coordinates are used homogeneously, we find that the matrices

$$\begin{bmatrix} \alpha & \beta & \gamma & \delta \\ \delta & -\gamma & \beta & -\alpha \\ \gamma & \delta & -\alpha & -\beta \\ -\beta & \alpha & \delta & -\gamma \end{bmatrix} \text{ and } \begin{bmatrix} x & t & -z & y \\ y & z & t & -x \\ z & -y & x & t \\ t & -x & -y & -z \end{bmatrix}$$

are orthogonal, and therefore their product is orthogonal. Now the rows of the first are obtained from $\alpha, \beta, \gamma, \delta$ by the operations *dd, ab, bc, ca,* and the columns of the second are obtained from x, y, z, t by operations on these letters analogous to *dd, ac, ba, cb,* and so the product has the same appearance as the multiplication table of these subgroups (p. 8) except as to the signs of some of its elements. The sign of each linear form in the product is the same as that of t, and so the product matrix

$$\begin{bmatrix} \alpha x+\beta y+\gamma z+\delta t, & -\delta x-\gamma y+\beta z+\alpha t, & \gamma x-\delta y-\alpha z+\beta t, & -\beta x+\alpha y-\delta z+\gamma t \\ \delta x-\gamma y+\beta z-\alpha t, & \alpha x-\beta y-\gamma z+\delta t, & \beta x+\alpha y-\delta z-\gamma t, & \gamma x+\delta y+\alpha z+\beta t \\ \gamma x+\delta y-\alpha z-\beta t, & \beta x+\alpha y+\delta z+\gamma t, & -\alpha x+\beta y-\gamma z+\delta t, & -\delta x+\gamma y+\beta z-\alpha t \\ -\beta x+\alpha y+\delta z-\gamma t, & \gamma x-\delta y+\alpha z-\beta t, & \delta x+\gamma y+\beta z+\alpha t, & -\alpha x-\beta y+\gamma z+\delta t \end{bmatrix}$$

is written

$$\begin{bmatrix} (dd) & (ac) & (ba) & (cb) \\ -(ab) & (da) & -(cc) & (bd) \\ -(bc) & (cd) & (db) & -(aa) \\ -(ca) & -(bb) & (ad) & (dc) \end{bmatrix}$$

and the linear forms occupy the same positions as the corresponding elements in the incidence diagram (p. 8). Hence in order to find a set of six, between any four of which a linear identity exists, we have to exclude from a row and a column the element common to both.

Many interesting geometrical theorems of fundamental importance for the configuration follow from the algebraic identities implied by the statement that *this matrix is orthogonal.* Write the matrix in the form

$$\begin{bmatrix} X_1 & X_2 & X_3 & X_4 \\ Y_1 & Y_2 & Y_3 & Y_4 \\ Z_1 & Z_2 & Z_3 & Z_4 \\ T_1 & T_2 & T_3 & T_4 \end{bmatrix}$$

then, taking any column, we have

$$X^2 + Y^2 + Z^2 + T^2 \equiv (\alpha^2 + \beta^2 + \gamma^2 + \delta^2)(x^2 + y^2 + z^2 + t^2),$$

showing that the four planes X, Y, Z, T are the faces of a tetrahedron self-polar with respect to the quadric $x^2 + y^2 + z^2 + t^2 = 0$, and correlatively the four points X, Y, Z, T are the corners of a self-polar tetrahedron which is, in fact, the same. Since a similar result follows from taking any row or column, out of the points and planes of the configuration *eight* tetrahedra* can be formed which are self-polar with respect to the quadric $x^2 + y^2 + z^2 + t^2 = 0$.

On subtracting the relations derived from the first row and column we get

$$X_2^2 + X_3^2 + X_4^2 - Y_1^2 - Z_1^2 - T_1^2 \equiv 0,$$

showing that the squares of the equations of six coplanar points are linearly connected, and therefore that the *six points lie on a conic*†. Correlatively six concurrent planes of the configuration touch a quadric cone.

Next, taking two columns we have identities such as

$$X_1X_2 + Y_1Y_2 + Z_1Z_2 + T_1T_2 \equiv 0,$$

showing that each of the tetrahedra

$$(X_1Y_1Z_1T_1) \text{ and } (X_2Y_2Z_2T_2)$$

is inscribed in and circumscribed about the other; so also the tetrahedra

$$(X_1Y_2Z_2T_1) \text{ and } (X_2Y_1Z_1T_2),$$
$$(X_2Y_1Z_2T_1) \text{ and } (X_1Y_2Z_1T_2),$$
$$(X_2Y_2Z_1T_1) \text{ and } (X_1Y_1Z_2T_2),$$

are similarly related. Further, the equations

$$X_1X_2 + Y_1Y_2 = 0,$$
$$Z_1Z_2 + T_1T_2 = 0$$

represent the same quadric surface, and so the lines $(X_1Y_1)(X_2Y_2)$ $(Z_1T_1)(Z_2T_2)$, shown in the first diagram, are generators of one

* Named after Rosenhain.

† Paul Serret, *Géométrie de Direction*, p. 132.

system, and (X_1Y_2) (X_2Y_1) (Z_1T_2) (Z_2T_1), shown in the second diagram, are generators of the other system of this quadric.

Another set of bilinear relations arises from the fact that complementary two-rowed subdeterminants of an orthogonal four-rowed matrix are equal (p. 27). Hence, for example,

$$X_1Y_2 - X_2Y_1 \equiv Z_3T_4 - Z_4T_3,$$

from which similar conclusions can be drawn.

Sets of eight *associated* points are represented by the bilinear relations of these two kinds; there are twelve of the former and eighteen of the latter, making *thirty* sets in all, and of the quadrics through any one set, four are plane-pairs, being planes of the configuration.

§ 17. THE TEN FUNDAMENTAL QUADRICS.

On substituting $x = \alpha$, $y = \beta$, $z = \gamma$, $t = \delta$, six of the linear forms vanish, namely, (ab) (ac) (bc) (ba) (ca) (cb), and the other ten become quadric functions of α, β, γ, δ which will be indicated by square brackets, thus $[dd] = \alpha^2 + \beta^2 + \gamma^2 + \delta^2$. These functions, equated to zero, represent the *fundamental quadric surfaces*, which play an important part in relation to the 16_6 configuration. After substitution, the orthogonal matrix of linear forms becomes a matrix of quadric functions

$$\begin{bmatrix} [dd] & 0 & 0 & 0 \\ 0 & [da] & -[cc] & [bd] \\ 0 & [cd] & [db] & -[aa] \\ 0 & -[bb] & [ad] & [dc] \end{bmatrix},$$

which is also orthogonal. Expressing this differently*, we have the result that an orthogonal three-rowed matrix is obtained when all the elements of the matrix

$$\begin{bmatrix} \alpha^2 - \beta^2 - \gamma^2 + \delta^2, & 2\alpha\beta - 2\gamma\delta, & 2\gamma\alpha + 2\beta\delta \\ 2\alpha\beta + 2\gamma\delta, & -\alpha^2 + \beta^2 - \gamma^2 + \delta^2, & 2\beta\gamma - 2\alpha\delta \\ 2\gamma\alpha - 2\beta\delta, & 2\beta\gamma + 2\alpha\delta, & -\alpha^2 - \beta^2 + \gamma^2 + \delta^2 \end{bmatrix}$$

are divided by $\alpha^2 + \beta^2 + \gamma^2 + \delta^2$.

* Rodrigues, *Liouville* (1840), v, 405; Darboux, *Comptes Rendus* (1881), XCII, 685.

These quadric surfaces are unchanged by the operations of the group, for every linear form is unchanged, except possibly as to sign, when the same operation is performed on both sets of symbols α, β, γ, δ, and x, y, z, t. It is on this fact that their importance chiefly depends.

The ten polars of any point are planes of the 16_6 configuration which is determined by that point, namely, those ten planes which do not pass through it; for $(\alpha, \beta, \gamma, \delta)$ is any point, and its polar plane with respect to (e.g.) the quadric

$$[aa] \equiv 2(\alpha\delta - \beta\gamma) = 0$$

is the plane

$$(aa) \equiv (\delta, -\gamma, -\beta, \alpha \between x, y, z, t) = 0,$$

and so on.

Hence the quadrics play fundamental and symmetrical parts in the configuration. In our nomenclature $[dd]$ has the peculiarity that pole and polar with respect to it have the same name, but this is only a convention and not essential. There is an incidence diagram and an orthogonal matrix of linear forms corresponding to each quadric, but some of the forms must have imaginary coefficients. There is a corresponding rearrangement of the ratios of the quadric functions to form a three-rowed orthogonal matrix.

If each of the fundamental quadric functions of α, β, γ, δ is multiplied by the same function of x, y, z, t, the sum of the ten products is

$$4(\alpha x + \beta y + \gamma z + \delta t)^2.$$

The ten quadrics are the only invariants of the second degree under the group of sixteen linear transformations.

Each of the ten quadrics corresponds to a partition of the operations ab, ac, bc, ba, ca, cb into two sets of three. The product of each set is the same, and gives the symbol for the quadric.

If the product of the operations p_1q_1 and p_2q_2 is p_3q_3 the point (p_1q_1) and the plane (p_2q_2) are pole and polar with respect to the quadric $[p_3q_3]$, and the four points (p_1q_1), (p_2q_2), (p_3q_3), (dd) are the corners of a tetrahedron self-polar with respect to $[dd]$.

§ 18. THE SIX FUNDAMENTAL COMPLEXES.

The six linear forms which vanish when $x = \alpha$, $y = \beta$, $z = \gamma$, $t = \delta$, are linear combinations of the six two-rowed determinants formed from the array

$$\begin{pmatrix} \alpha & \beta & \gamma & \delta \\ x & y & z & t \end{pmatrix},$$

and these are the Plücker coordinates of the line joining the points $(\alpha, \beta, \gamma, \delta)$ and (x, y, z, t). Write

$$p_{14} = \alpha t - \delta x, \quad p_{24} = \beta t - \delta y, \quad p_{34} = \gamma t - \delta z,$$
$$p_{23} = \beta z - \gamma y, \quad p_{31} = \gamma x - \alpha z, \quad p_{12} = \alpha y - \beta x.$$

Then

$$(ab) = (\delta, \gamma, \beta, \alpha \between -x, \ y, -z, t) = p_{14} - p_{23},$$
$$(ac) = (\delta, \gamma, \beta, \alpha \between -x, -y, \ z, t) = p_{14} + p_{23},$$
$$(bc) = (\gamma, \delta, \alpha, \beta \between -x, -y, \ z, t) = p_{24} - p_{31},$$
$$(ba) = (\gamma, \delta, \alpha, \beta \between \ x, -y, -z, t) = p_{24} + p_{31},$$
$$(ca) = (\beta, \alpha, \delta, \gamma \between \ x, -y, -z, t) = p_{34} - p_{12},$$
$$(cb) = (\beta, \alpha, \delta, \gamma \between -x, \ y, -z, t) = p_{34} + p_{12}.$$

On equating these to zero we get six *fundamental linear complexes*, and the null-planes of $(\alpha, \beta, \gamma, \delta)$ are those six planes of the configuration which pass through it. In the next chapter will be found a detailed account of this system of complexes. As their name implies, they are of fundamental importance in the theory of the 16_6 configuration, and are unchanged when the same operation of the group is performed upon $(\alpha, \beta, \gamma, \delta)$ and (x, y, z, t).

It is to be noticed that the ten quadrics and six complexes are determined by the coordinates alone, and that then the points of the configuration are the ten poles and six null-points of an arbitrary plane, and the planes of the configuration are the polars and null-planes of any one of its points.

The coefficients in the linear identities connecting these six forms are the fundamental quadric functions of $\alpha, \beta, \gamma, \delta$; we easily find, in particular,

$$[dd](ab) \equiv \ [da](ac) - [cc](ba) + [bd](cb),$$
$$[dd](bc) \equiv \ [cd](ac) + [db](ba) - [aa](cb),$$
$$[dd](ca) \equiv -[bb](ac) + [ad](ba) + [dc](cb),$$

and the identity

$$(ab)^2 + (bc)^2 + (ca)^2 \equiv (ac)^2 + (ba)^2 + (cb)^2$$

shows once more that the nine quadrics on the right, when divided by $[dd]$, are the elements of an orthogonal matrix.

§ 19. IRRATIONAL EQUATIONS OF KUMMER'S SURFACE.

Corresponding to any identical relation among the planes of the configuration of the form

$$Z_1 T_1 + Z_2 T_2 + Z_3 T_3 + Z_4 T_4 = 0$$

and any three constants λ, μ, ν satisfying

$$\lambda+\mu+\nu=0,$$

the four equations

$$\sqrt{\lambda Z_2 T_2}+\sqrt{\mu Z_3 T_3}+\sqrt{\nu Z_4 T_4}=0,$$
$$\sqrt{\lambda Z_1 T_1}+\sqrt{\mu Z_4 T_4}+\sqrt{\nu Z_3 T_3}=0,$$
$$\sqrt{\lambda Z_4 T_4}+\sqrt{\mu Z_1 T_1}+\sqrt{\nu Z_2 T_2}=0,$$
$$\sqrt{\lambda Z_3 T_3}+\sqrt{\mu Z_2 T_2}+\sqrt{\nu Z_1 T_1}=0$$

represent the same quartic surface, having the eight planes $Z_1=0 \ldots T_4=0$ for tropes and the eight points of the configuration in which they meet by fours for nodes. From the way in which the elements of the configuration are interchanged by the group we see that there are eight transformations of x, y, z, t which leave this surface unaltered. If we choose $\lambda:\mu:\nu$ so that the surface pass through an additional point of the configuration, it will then necessarily pass through all the remaining points, and must therefore coincide with the Kummer surface associated with the configuration.

Suppose that $(\alpha, \beta, \gamma, \delta)$ is not one of the points through which the surface passes for arbitrary values of $\lambda:\mu:\nu$ and let the result of substituting $(\alpha, \beta, \gamma, \delta)$ for (x, y, z, t) be indicated by square brackets; further suppose that $[Z_1]\equiv 0 \equiv [T_1]$. Then the surface will pass through $(\alpha, \beta, \gamma, \delta)$ if

$$\lambda/[Z_2 T_2]=\mu/[Z_3 T_3]=\nu/[Z_4 T_4],$$

agreeing with $\lambda+\mu+\nu=0$ in virtue of the relations among the fundamental quadrics.

Hence Kummer's surface can be written in the form

$$\sqrt{[Z_2 T_2]\, Z_2 T_2}+\sqrt{[Z_3 T_3]\, Z_3 T_3}+\sqrt{[Z_4 T_4]\, Z_4 T_4}=0,$$

and seven other forms can be deduced from this by the group of linear transformations which leave the surface unaltered. This set of eight equations is connected with two complementary groups of eight associated points, and so the total number of equations is $15 \cdot 8=120$. One equation of each set is included in the system of fifteen equations formed by the following rule. Multiply each linear form by the corresponding quadric, that is, its value at $(\alpha, \beta, \gamma, \delta)$, and put the square root of the product in place of the linear form in the orthogonal matrix; then the quadratic relations, not identities, among the elements of the matrix so formed, which are satisfied when it is orthogonal, are equations of Kummer's surface.

In order to express these equations in the notation of § 10, we arrange the sixteen expressions $x_1 \ldots x_{123} \ldots$, multiplied by suitable constants, in an orthogonal matrix, and then compare the elements with the linear forms. Let

$$f(k) \equiv (k-k_1)(k-k_2)(k-k_3)(k-k_4)(k-k_5)(k-k_6),$$

$$\xi_1 \sqrt{-f'(k_1)} = u_1 v_1 = x_1, \qquad \xi_2 \sqrt{f'(k_2)} = u_2 v_2 = x_2, \text{ etc.};$$

further let

$$c_{123}^{-2} = -c_{456}^{-2}$$
$$= (k_1-k_4)(k_1-k_5)(k_1-k_6)(k_2-k_4)(k_2-k_5)(k_2-k_6)(k_3-k_4)(k_3-k_5)(k_3-k_6),$$

$$\xi_{123}/c_{123} = (\sqrt{u_1 u_2 u_3 v_4 v_5 v_6} + \sqrt{v_1 v_2 v_3 u_4 u_5 u_6})^2/(u-v)^2 = x_{123};$$

then

$$\begin{bmatrix} \xi_{135} & \xi_2 & \xi_4 & \xi_6 \\ \xi_1 & \xi_{146} & \xi_{162} & \xi_{124} \\ \xi_3 & \xi_{346} & \xi_{362} & \xi_{324} \\ \xi_5 & \xi_{546} & \xi_{562} & \xi_{524} \end{bmatrix}$$

is orthogonal, and if $k_1 < k_2 < k_3 < k_4 < k_5 < k_6$ all the elements are real. Further the matrix

$$\begin{bmatrix} \sqrt{c_{135}\xi_{135}} & 0 & 0 & 0 \\ 0 & \sqrt{c_{146}\xi_{146}} & \sqrt{c_{162}\xi_{162}} & \sqrt{c_{124}\xi_{124}} \\ 0 & \sqrt{c_{346}\xi_{346}} & \sqrt{c_{362}\xi_{362}} & \sqrt{c_{324}\xi_{324}} \\ 0 & \sqrt{c_{546}\xi_{546}} & \sqrt{c_{562}\xi_{562}} & \sqrt{c_{524}\xi_{524}} \end{bmatrix}$$

is also orthogonal, showing that c_{135}, etc. correspond to fundamental quadrics*.

* Cf. *Math. Annalen*, Staude, xxiv, 281; Klein, xxvii, 431; Bolza, xxx, 478.

CHAPTER IV.

LINE GEOMETRY.

§ 20. POLAR LINES.

It is from the point of view of line geometry that the Kummer configuration receives its most natural and symmetrical development, for the complete reciprocity exhibited by the configuration indicates that its properties are based on a system of geometry in which the *line* is taken to be the fundamental element.

A line is capable of satisfying four conditions (though some conditions must be reckoned as two-fold and some three-fold), a fact which may be expressed by saying that there are ∞^4 lines in space. A *complex* consists of the ∞^3 lines satisfying one condition, and is *algebraic* if on adding any three-fold condition the lines are reduced to a finite number. If the three-fold condition is that of belonging to a given plane pencil, this finite number is called the *degree* of the complex.

In this chapter only *linear* complexes will be considered, namely those in which only one line passes through a given point and at the same time lies in a given plane. It follows from this definition that all the lines belonging to a linear complex which pass through a given point lie in a plane, its *null-plane*, and all the lines in a given plane pass through a point, its *null-point*. From these facts all the properties of the familiar null-system can be deduced.

The chief property that will be used is that of *polar lines**. We regard a linear complex as establishing a *correlation* between null-point and null-plane and an involutory correspondence between polar lines. Introducing a symbol S to denote the correspondence we may write

$$S(P)=\pi, \quad S(\pi)=P, \quad S^2=1,$$

* In works on Statics these are called *conjugate lines* (Routh), and *reciprocal lines* (Minchin).

P being any point and π its null-plane, and the symbol 1 denoting the identical transformation. Let P' be any other point and π' its null-plane, then if x is the line PP' and y the line $\pi\pi'$ we may write

$$S(x) = y, \quad S(y) = x$$

and x, y are by definition polar lines.

§ 21. APOLAR COMPLEXES.

In general two such correspondences S, T are not commutative; when they are, that is when

$$ST = TS,$$

the complexes will be said to be *apolar** and ST or TS may be taken to be the symbol of an involutory point-point and plane-plane linear transformation or *collineation*†; for any plane π has two null-points $P = S(\pi)$ and $Q = T(\pi)$, and

$$TS(P) = TS^2(\pi) = T(\pi) = Q,$$
$$ST(Q) = ST^2(\pi) = S(\pi) = P,$$

so that when $ST = TS$ the correspondence between P and Q is involutory. The line PQ is a common ray of the two complexes and the transformation ST or TS determines an involution on it.

Let x and y be any pair of polar lines with respect to the first complex, then

$$S(x) = y,$$
$$TS(x) = T(y),$$
$$S\{T(x)\} = T(y),$$

and the transformation T does not destroy the relation of polarity with respect to the first complex. Further, if x is a ray of the second complex,

$$T(x) = x,$$
$$T(y) = S(x) = y,$$

and y is also a ray. The relation between apolar complexes is

* The customary terms for this relationship are *in involution* (Klein) and *reciprocal* (Ball); the former is awkward and the latter suggests a false analogy. Two quantics are apolar when their transvectant of highest index vanishes identically, and this can be interpreted for complexes of any degree.

† Any two lines determine such a collineation; "geschaart-involutorisches System," Reye, *Geometrie der Lage*, II, 17; "système involutif gauche," Reye-Chemin, *Geometrie der Position*, 145; "windschiefe Involution," Sturm, *Liniengeometrie*, I, 70, 115.

therefore such that the polars with respect to one complex of the rays of the other are also rays of the other.

The assemblage of ∞^2 rays common to two complexes is called a *congruence*; there are two common polar lines called its *directrices*, which meet every ray. In the present case the directrices satisfy the relation

$$S(x) = T(x),$$

whence $$TS(x) = x,$$

and the common polar lines correspond to themselves in the transformation ST or TS, and therefore cut any common ray in the double points of the involution on it. Hence any two corresponding lines and the two directrices form a set of four harmonic generators of a *regulus*.

An important property of apolar complexes is that they lead to finite groups of transformations; that is, the repetition of the operation of taking the null-point (or null-plane) of a plane (or point) leads to a finite number of points and planes. Thus in the case of two complexes if S_1 and S_2 are the correlations determined by them, they determine a group of collineations containing two members 1 and $S_1 S_2$, so that an arbitrary point or plane gives rise to a figure of two points and two planes.

§ 22. GROUPS OF THREE AND FOUR APOLAR COMPLEXES.

Three mutually apolar complexes determine three correlations S_1, S_2, S_3, which give rise to three collineations,

$$T_1 = S_2 S_3, \quad T_2 = S_3 S_1, \quad T_3 = S_1 S_2,$$

forming a group $(1, T_1, T_2, T_3)$ whose multiplication table is

	T_1	T_2	T_3
T_1	1	T_3	T_2
T_2	T_3	1	T_1
T_3	T_2	T_1	1

similar to that of § 2. Hence an arbitrary point P gives rise to three concurrent planes $S_1(P)$, $S_2(P)$, $S_3(P)$ and three points $T_1(P)$, $T_2(P)$, $T_3(P)$ lying on their lines of intersection. Since

$$S_1 T_1(P) = S_2 T_2(P) = S_3 T_3(P) = S_1 S_2 S_3(P)$$

the plane containing $T_1(P)$, $T_2(P)$, $T_3(P)$ is the null-plane of each in the corresponding complex. We have therefore altogether four points and four planes forming a tetrahedron such that any three faces are the null-planes of their common corner and any three corners are the null-points of the face containing them in the three complexes.

The rays common to the three complexes are the generators of a regulus and the three pairs of directrices of the three congruences belong to the complementary regulus. Hence the points P and $S_1S_2(P)$ being harmonically separated by the directrices of the congruence (S_1S_2) are conjugate points with respect to the quadric on which these reguli lie; it follows that the tetrahedron is self-polar.

Four mutually apolar complexes give four correlations S_1, S_2, S_3, S_4, and an arbitrary point gives rise to points and planes which may be denoted by the symbols of the corresponding transformations. Thus from an arbitrary point 1 we derive

four planes S_1, S_2, S_3, S_4,
six points $S_1S_2, S_1S_3\ldots$,
four planes $S_1S_2S_3\ldots$,
and one point $S_1S_2S_3S_4$,

making eight points and eight planes altogether. If we arrange the points thus

$$\begin{array}{cccc} 1 & S_2S_3 & S_3S_1 & S_1S_2 \\ S_1S_2S_3S_4 & S_1S_4 & S_2S_4 & S_3S_4 \end{array}$$

the first row contains the corners of a tetrahedron whose faces are

$$S_1S_2S_3 \quad S_1 \quad S_2 \quad S_3$$

and each of these planes contains one of the points in the second row, which is its null-point in the complex S_4. Thus the configuration can be regarded in four ways as a pair of circumscribed and inscribed tetrahedra.

§ 23. SIX APOLAR COMPLEXES.

The existence of six mutually apolar complexes* depends on the well-known fact that the condition of being in involution with a given complex is a one-fold condition. Hence we may take the first complex arbitrarily, the second from the ∞^4 complexes

* Klein, *Math. Annalen* (1870), II, 198; Sturm, *Liniengeometrie* (1892), I, 234 Koenigs, *La Géométrie Réglée* (1895), p. 92; Ball, *Theory of Screws* (1900), p. 33.

apolar to the first and so on, the last being uniquely determined by the preceding five.

Assuming such a set, we may denote the complexes by the symbols 1, 2, 3, 4, 5, 6. Taken in pairs they determine *fifteen* congruences (12), etc., and taken in threes they determine *twenty* reguli (123), etc. We shall first examine the relations of special lines connected with these complexes, and afterwards consider configurations derived from an arbitrary point, line, or plane by means of the associated transformations.

Any four of the complexes, say 1, 2, 3, 4, have two common rays; since, by the hypothesis of apolarity, the polar of either of these with respect to 5 or 6 must belong to 1, 2, 3 and 4, it must be the other common ray. Hence these lines, being a common pair of polar lines with respect to 5 and 6, are the directrices of the congruence (56).

The directrices of (12) and (13) do not meet, for they belong to the regulus (456); and since the latter pair are rays of 2 and polar lines with respect to 1, they correspond in the collineation determined by (12) and therefore separate the former pair harmonically (p. 39). In other words, the four directrices of (12) and (13) cut the generators of (123) in harmonic ranges. Similar reasoning shows that any two pairs of the directrices (12), (23), (31) are harmonically conjugate.

The directrices of (12) cut those of (34) and hence the directrices of (12), (34), (56) are the edges of a tetrahedron. There are *fifteen* of these *fundamental tetrahedra*, which may be denoted by the symbols (12, 34, 56), etc.

§ 24. TEN FUNDAMENTAL QUADRICS.

Since the directrices (12), (23), (31) are rays of 4, 5, 6, it follows that the reguli (123), (456) lie on the same quadric, which may be called (123, 456). These are the ten *fundamental quadrics*.

The quadrics (123, 456) and (124, 356) intersect in the rays common to 1, 2, 3, 4, and in those common to 3, 4, 5, 6. Hence they have the quadrilateral of directrices (56), (12) common. It follows that the directrices (34), being the diagonals of the quadrilateral, are a pair of polar lines with respect to both quadrics. Thus any two fundamental quadrics have contact at four points, which are corners of a fundamental tetrahedron, and at each of the sixty corners three pairs of quadrics have contact.

The tetrahedron (14, 25, 36) is self-polar with respect to the quadric (123, 456). Hence of the fifteen tetrahedra nine are inscribed in any one of the ten quadrics, and the remaining six are self-polar, and of the ten quadrics six are circumscribed about any one tetrahedron which is self-polar with respect to the remaining four.

Each of the four quadrics

(123, 456)
(126, 453)
(153, 246)
(156, 243)

with respect to which (14, 25, 36) is self-polar, is its own reciprocal with respect to each of the other three; for the first is completely determined by the four directrices (12) and (13) whose polars with respect to the second quadric are the same four lines.

§ 25. KLEIN'S 60_{15} CONFIGURATION.

The fifteen fundamental tetrahedra have *thirty* edges; each edge is common to three tetrahedra, thus a directrix of (12) is common to (12, 34, 56), (12, 35, 46), (12, 45, 36). Of the *sixty* corners, six lie on each edge and, as has been shown, are arranged as three pairs forming three harmonic ranges. So also of the *sixty* faces, six pass through an edge and the pair belonging to one tetrahedron are harmonically conjugate with respect to the pair belonging to either of the other two tetrahedra having the same edge. In each face are three edges, each containing six corners, giving *fifteen* corners in that face, and similarly *fifteen* faces pass through one corner.

We may distinguish the directrices of the congruence (12) by the symbols 12 and 21, and make the convention that sets of three lines obtained by an even number of interchanges of figures from 12, 34, 56 shall be coplanar. Then those obtained by an odd number of interchanges will be concurrent. Thus, for example, the rows

12	34	56
46	25	13
35	61	24

are coplanar and the columns concurrent. Hence the three planes intersect in the line of collinearity of the three points. An even number of interchanges applied to the symbols common to two rows and two columns leaves the rows coplanar and the columns concurrent as before, thus the table of lines

21 43 56
64 52 13
35 61 24

possesses properties similar to those of the former table, and so on. Hence four pairs of corners, one from each of the tetrahedra (12, 46, 35) and (34, 25, 16), are collinear with the point 56, 13, 24, and the same is true for the other corners of (56, 13, 42) so that the two former tetrahedra are in fourfold perspective. Hence the three tetrahedra represented by the columns of the table belong to a *desmic system* (§ 1) and the rows represent the other desmic system which is formed with the same edges.

Again the following three sets of lines are similarly related,

12	34	56	12	34	56	12	34	56
46	15	32	54	16	32	54	26	13
35	26	14	36	52	14	36	15	24

showing that there are four desmic systems containing the same one tetrahedron (12, 34, 56). Hence through any one corner of the configuration pass sixteen lines containing two other corners, and in each face lie sixteen lines through which pass two other faces. From the desmic properties it follows that the assemblage of these lines is the same in each case, and this number is $60 . 16/3 = 320$.

It is possible in six different ways, corresponding to the six pairs of different cyclical arrangements of five figures (§ 6), to select a set of five tetrahedra including all thirty directrices among their edges; consequently no two tetrahedra of the same set have a common edge. One such set occupies the first column of the following table*:

$(ab)=(12, 34, 56)$	$(bc)=(16, 24, 35)$	$(ce)=(14, 23, 56)$
$(ac)=(13, 25, 46)$	$(bd)=(15, 23, 46)$	$(cf)=(15, 26, 34)$
$(ad)=(14, 26, 35)$	$(be)=(13, 26, 45)$	$(de)=(16, 25, 34)$
$(ae)=(15, 24, 36)$	$(bf)=(14, 25, 36)$	$(df)=(13, 24, 56)$
$(af)=(16, 23, 45)$	$(cd)=(12, 36, 45)$	$(ef)=(12, 35, 46)$

* Richmond, *Quarterly Journal*, XXXIV, 124.

and the members of any one set are distinguished by two-letter symbols having one letter common. Two tetrahedra such as (ab) and (ac) from the same set belong to a desmic system of which the third member is (bc), not belonging to the same set. Thus we have *twenty* desmic systems corresponding to the combinations of the six letters a, b, c, d, e, f three at a time. Further, three tetrahedra such as (ab) (cd) (ef) have two edges common.

§ 26. KUMMER'S 16_6 CONFIGURATION.

We now turn to configurations containing arbitrary elements. Using the symbols 1, 2, 3, 4, 5, 6 to represent the permutable correlations determined by the six complexes, we obtain from any point P the points

$$12P,\quad 13P \ldots\quad 1234P,\quad 1235P \ldots\quad 123456P.$$

Call the last point Q, then $12Q = 3456P$ and the points are

$$P,\quad 12P,\quad 13P \ldots\quad 56P,$$
$$Q,\quad 12Q,\quad 13Q \ldots\quad 56Q.$$

Now it was proved in the case of three complexes that the points $P, 23P, 31P, 12P$ form a tetrahedron self-polar with respect to the quadric (123); similarly the tetrahedron Q, $56Q$, $64Q$, $45Q$ is self-polar with respect to the quadric (456). But these quadrics are the same, and, further, it was proved in the case of four complexes that the plane $23P, 31P, 12P$ contains the point $1234P$, or $56Q$. Similarly this plane contains also $64Q$ and $45Q$, so that the two tetrahedra have a common face, and therefore P and Q coincide with the pole of this face with respect to the quadric (123, 456).

We have therefore derived from an arbitrary point a configuration consisting of sixteen points and sixteen planes, which is, in fact, Kummer's 16_6 configuration. The planes consist of the six null-planes and ten polar planes of P. That the six points in the polar plane with respect to (123, 456) lie on a conic follows from the fact that the triangles 12, 23, 31 and 45, 56, 64 are self-polar with respect to the section by their common plane.

This method leads to the same nomenclature as was used in § 9. An arbitrary point 0 has six null-planes 1, 2, 3, 4, 5, 6 named after the corresponding complexes. The null-points of these planes lie on their common lines, and are the remaining fifteen points, named 12 ... 56. The remaining ten planes are the polars of 0 with respect to the fundamental quadrics, and accordingly receive the same names. The nomenclature is thus interpreted by regarding the name of each element as the symbol of operation deriving it from an arbitrary point 0. The symbol 0 itself must therefore denote the identical operation and the other operations 1, 2, 3, 4, 5, 6 obey the laws expressed symbolically by

$$0 = 11 = 22 = 33 = 44 = 55 = 66 = 123456.$$

The diagram (p. 17)

0	46	62	24
35	12	14	16
51	32	34	36
13	52	54	56

showing which points are coplanar is now seen to be also a multiplication table for the sub-groups (0, 35, 51, 13) and (0, 46, 62, 24) each of four collineations. The corresponding diagram for planes may be written down from the known laws of incidence, and is

135	2	4	6
1	146	162	124
3	346	362	324
5	546	562	524

but we see that it can be obtained from the former by operating on each element with 135*.

§ 27. LINE COORDINATES.

The ideas and methods hitherto employed in the present chapter are purely geometrical, and there is no doubt that every theorem can be deduced in this way; but in many cases the proofs are artificial and tedious and an analytical method is more direct and throws more light on the true lines of reasoning.

The theory of apolar linear complexes may be investigated with exceptional elegance by means of line coordinates†. With generalised coordinates x_1, x_2, x_3, x_4, x_5, x_6 let $\omega(x) = 0$ be the

* For detailed elaboration of these configurations the following references may be consulted: Caporali, *Memorie Lincei* (1878), ser. 3, II, 3; Stephanos, *Darboux Bulletin* (1879), sér. 2, III, 424; Hess, *Nova Acta*, Halle (1891), LV, 96; Martinetti, *Rendiconti Palermo* (1902), XVI, 196.

† It is assumed that the reader is acquainted with the subject-matter of the first two chapters of Prof. Jessop's *Treatise on the Line Complex*. See also Koenigs, *La Géométrie Réglée*.

quadratic relation among them and let $\Omega(a)=0$ be the condition that the linear complex $\Sigma a_s x_s = 0$ may be special; then the matrices of the coefficients in ω and Ω are inverse*. The complexes $\Sigma a_s x_s = 0$ and $\Sigma b_s x_s = 0$ are apolar if

$$\Sigma a_s \partial\Omega(b)/\partial b_s = 0.$$

When the two complexes are taken to be the coordinate complexes $x_1 = 0$ and $x_2 = 0$, the condition for apolarity becomes

$$\partial^2\Omega(a)/\partial a_1 \partial a_2 = 0,$$

in other words, the term in $a_1 a_2$ must be absent from $\Omega(a)$. Hence the problem of finding six mutually apolar complexes is the same as that of expressing the quadratic form $\Omega(a)$ as the sum of six squares. We shall suppose that this has been done; then the quadratic form $\omega(x)$ is at the same time reduced to the sum of six squares, and by taking suitable multiples of the coordinates we can make

$$\omega(x) = \Omega(x) = x_1^2 + x_2^2 + x_3^2 + x_4^2 + x_5^2 + x_6^2.$$

By using these coordinates calculations are much simplified, though the meaning of general results is liable to become obscured through the identity of ω and Ω. A particular example of these coordinates is

$$x_1 = p_{14} - p_{23}, \quad x_3 = p_{24} - p_{31}, \quad x_5 = p_{34} - p_{12},$$
$$ix_2 = p_{14} + p_{23}, \quad ix_4 = p_{24} + p_{31}, \quad ix_6 = p_{34} + p_{12},$$

where it will be noticed that all the coordinate complexes are real, but three of the coordinates are imaginary. On referring to § 16 we see that the coordinate complexes are the fundamental complexes employed in the construction of the 16_6 configuration.

The lines $(x_1, x_2, x_3, x_4, x_5, x_6)$ and $(-x_1, x_2, x_3, x_4, x_5, x_6)$ are polar with respect to the complex $x_1 = 0$. Hence the transformation of lines associated with each complex consists in changing the sign of the corresponding coordinate. Taking only the first three complexes, a line (x) gives rise to a set of eight lines $(\pm x_1, \pm x_2, \pm x_3, x_4, x_5, x_6)$. In order to deal with points and planes we must suppose (x) to describe a sheaf of lines through a point P, then $(-x_1, x_2, x_3, x_4, x_5, x_6)$ describes a plane field $S_1(P)$, the null-plane of P in the first complex†, and $(-x_1, -x_2, x_3, x_4, x_5, x_6)$ describes a sheaf whose vertex is $S_2S_1(P)$ or $T_3(P)$, and so on. This method expresses the operations of the group $(1, T_1, T_2, T_3)$ in a form which brings out clearly the comparison with the group of reflexions (§ 2).

* A simple example in matrix notation.

† Compare § 22.

§ 28. FUNDAMENTAL QUADRICS.

The condition of intersection of two lines (y) and (z) being

$$y_1z_1 + y_2z_2 + y_3z_3 + y_4z_4 + y_5z_5 + y_6z_6 = 0$$

it is evident that the lines

$$(y_1, y_2, y_3, 0, 0, 0), \quad (0, 0, 0, z_4, z_5, z_6)$$

intersect; this proves that the lines common to $x_1 = x_2 = x_3 = 0$ and those common to $x_4 = x_5 = x_6 = 0$ are the two systems of generators of the same quadric surface (123, 456). The general tangent line to this surface is a ray of the plane pencil determined by two intersecting generators and has coordinates

$$\lambda y_1, \lambda y_2, \lambda y_3, \mu z_4, \mu z_5, \mu z_6$$

where $$y_1^2 + y_2^2 + y_3^2 = 0 = z_4^2 + z_5^2 + z_6^2.$$

Hence any tangent (x) satisfies the equivalent equations

$$x_1^2 + x_2^2 + x_3^2 = 0, \quad x_4^2 + x_5^2 + x_6^2 = 0,$$

either of which is the *line equation* of the quadric (123, 456).

The equation $x_1^2 + x_2^2 + x_3^2 = 0$ implies that the three poles of any plane with respect to the three complexes $x_1 = 0$, $x_2 = 0$, $x_3 = 0$ form a triangle self-polar with respect to the section of the quadric (123, 456); for, if we substitute for the x_s bilinear expressions in terms of the coordinates of two planes and regard one of these planes as fixed, $x_1 = 0$, $x_2 = 0$, $x_3 = 0$ become the tangential equations of three points and $x_1^2 + x_2^2 + x_3^2 = 0$ that of a conic with respect to which these points are mutually conjugate.

In a similar way the equation

$$x_1^2 + x_2^2 + x_3^2 + x_4^2 + x_5^2 + x_6^2 = 0$$

shows that the six poles of any plane lie on a conic, for it expresses that the squares of the tangential equations of these points are linearly connected, which is a necessary and sufficient condition (p. 31).

In terms of *real* coordinates $z_1 = p_{14} - p_{23}$, $z_2 = p_{14} + p_{23}$, etc., the equation may be written

$$z_1^2 + z_3^2 + z_5^2 = z_2^2 + z_4^2 + z_6^2,$$

showing that the triangle of poles 246 is obtained from the triangle 135 by a transformation which may be regarded as a *rigid rotation* in an "elliptic" plane. From this we infer that the points 135 occur *alternately* with the points 246 on the conic. If the quadric (135, 246) is regarded as the "absolute" of an elliptic space,

(z_1, z_3, z_5) and (z_2, z_4, z_6) are the two Clifford parallels through a corner of the tetrahedron of reference to any line*.

It is easily seen, by eliminating one set of point coordinates from the bilinear expressions for x_s in terms of two points, that any four line coordinates are connected by a linear relation in which the coefficients are quadratic in point coordinates; let one such relation be

$$Q_{234}x_1 + Q_{134}x_2 + Q_{124}x_3 + Q_{123}x_4 = 0,$$

then the equation $Q_{123} = 0$ evidently represents the regulus of rays common to $x_1 = x_2 = x_3 = 0$. Similarly we have relations

$$Q_{235}x_1 + Q_{135}x_2 + Q_{125}x_3 + Q_{123}x_5 = 0,$$

$$Q_{236}x_1 + Q_{136}x_2 + Q_{126}x_3 + Q_{123}x_6 = 0.$$

Then the equation $\Sigma x_s^2 = 0$, in which x_1, x_2, x_3 are regarded as arbitrary, shows that the coefficients of x_1, x_2, x_3 in the preceding relations, when divided by iQ_{123}, form an orthogonal matrix (§ 17).

§ 29. FUNDAMENTAL TETRAHEDRA.

Rays common to the four complexes $x_1 = 0, x_2 = 0, x_3 = 0, x_4 = 0$ satisfy

$$x_1^2 + x_2^2 = 0.$$

There are therefore two common rays

$$(1, i, 0, 0, 0, 0) \text{ and } (1, -i, 0, 0, 0, 0),$$

which are seen to be polar lines with respect to both of the complexes $x_1 = 0$, $x_2 = 0$, and are therefore the directrices of the congruence (12). Similarly the coordinates of all the other directrices may be found.

The edges of the tetrahedron (12, 34, 56) have *line equations*

$$x_1 \pm ix_2 = 0, \quad x_3 \pm ix_4 = 0, \quad x_5 \pm ix_6 = 0.$$

If this tetrahedron is real and is taken for reference the equation of any edge is expressed by the vanishing of the corresponding Plücker line coordinate. By taking suitable multiples of the point coordinates we may arrange that

$$x_1 = p_{14} - p_{23}, \quad x_3 = p_{24} - p_{31}, \quad x_5 = p_{34} - p_{12},$$

$$ix_2 = p_{14} + p_{23}, \quad ix_4 = p_{24} + p_{31}, \quad ix_6 = p_{34} + p_{12},$$

so that the particular example of p. 46 is really of general significance. In this way the Plücker coordinates of all the directrices

* Whitehead, *Universal Algebra*, p. 405.

may be found and thence their intersections, forming the points of Klein's configuration.

There are two tetrahedra (12, 35, 46) and (12, 36, 45) which have the edges (12) in common with the tetrahedron (12, 34, 56) of reference, and it has been proved (§ 23) that the corners on a common edge form three harmonic ranges. It is easily found that the corners of (12, 35, 46) are $(1, 0, 0, \pm i)$, $(0, 1, \pm i, 0)$, and those of (12, 36, 45) are $(1, 0, 0, \pm 1)$, $(0, 1, \pm 1, 0)$. In this way, by taking different pairs of opposite edges of reference, six tetrahedra are found. The remaining eight form with (12, 34, 56) four desmic systems, and it is therefore sufficient to give one corner in each system (§ 1). These points are $(1, 1, 1, 1)$, $(1, i, i, 1)$, $(i, 1, i, 1)$, $(i, i, 1, 1)$, and the corners of any tetrahedron are obtained from these by changing the signs of an odd or even number of coordinates. For example (13, 25, 46) is $(-i, 1, i, 1)$, $(i, -1, i, 1)$, $(i, 1, -i, 1)$, $(i, 1, i, -1)$.

Klein's configuration can be constructed from a single tetrahedron as follows, and the process verifies that the number of arbitrary constants is the same as for six apolar complexes, namely $5+4+3+2+1+0=15$.

We recall the fact that if $u=0$ and $v=0$ are the equations of any two points then the general pair of harmonic conjugates is given by $u^2-\lambda^2v^2=0$, and the condition that another pair, $u^2-\mu^2v^2=0$, may be harmonically conjugate with the preceding pair is $\lambda^2+\mu^2=0$.

Take one tetrahedron arbitrarily for reference (twelve constants). Three of those which have two edges in common with it are in the first instance

$$\left.\begin{matrix}(a, & 0, & 0, & \pm 1)\\ (0, & a', & \pm 1, & 0\ )\end{matrix}\right\} \quad \left.\begin{matrix}(\ 0, & b, & 0, & \pm 1)\\ (\pm 1, & 0, & b', & 0\ )\end{matrix}\right\} \quad \left.\begin{matrix}(0, & 0, & c, & \pm 1)\\ (c', & \pm 1, & 0, & 0\ )\end{matrix}\right\}$$

and the new constants must be chosen so that the edges intersect. The points $(\lambda a, \lambda' a', \lambda', \lambda)$ and $(\mu', \mu b, \mu' b', \mu)$ can by proper choice of $\lambda, \lambda', \mu, \mu'$ be made the same if $aa'=b/b'$, and so on; thus $a'=b/c$, $b'=c/a$, $c'=a/b$. By taking new multiples of the coordinates and thereby absorbing three arbitrary constants we may put $a=b=c=1$, and then $a'=b'=c'=1$. The preceding three tetrahedra are now

$$\left.\begin{matrix}(1, & 0, & 0, & \pm 1)\\ (0, & 1, & \pm 1, & 0\ )\end{matrix}\right\} \quad \left.\begin{matrix}(\ 0, & 1, & 0, & \pm 1)\\ (\pm 1, & 0, & 1, & 0\ )\end{matrix}\right\} \quad \left.\begin{matrix}(0, & 0, & 1, & \pm 1)\\ (1, & \pm 1, & 0, & 0\ )\end{matrix}\right\}$$

and the remaining three having pairs of edges in common with the tetrahedron of reference are

$$\left.\begin{matrix}(1, 0, 0, \pm i)\\(0, 1, \pm i, 0)\end{matrix}\right\} \quad \left.\begin{matrix}(0, 1, 0, \pm i)\\(\pm i, 0, 1, 0)\end{matrix}\right\} \quad \left.\begin{matrix}(0, 0, 1, \pm i)\\(1, \pm i, 0, 0)\end{matrix}\right\}$$

The rest of the configuration is now completely determined by the intersections of the edges of these seven tetrahedra.

The fifteen tetrahedra play symmetrical parts in the configuration and each belongs to four desmic systems. It is possible to represent five tetrahedra by products each of four linear factors, such that the ten differences of these products are also products of four factors and represent the remaining tetrahedra of the system. In this way the two-letter nomenclature of p. 43 may be derived. See the second example on p.

CHAPTER V.

THE QUADRATIC COMPLEX AND CONGRUENCE.

§ 30. OUTLINE OF THE GEOMETRICAL THEORY.

We have seen in the preceding chapter how the configuration of nodes and tropes of a Kummer surface arises naturally in elementary line geometry from the consideration of a set of apolar linear complexes. It will now be shown how the surface itself occurs as the *singular surface* of a quadratic complex of general character, which is self-polar with respect to each of the former set.

In the present section the leading ideas in the theory of quadratic complexes are presented in outline, and the reader is referred to existing treatises for proofs and fuller accounts*. The geometrical method will be followed up to a certain point, after which it is more advantageous to adopt the treatment by coordinates.

The rays of a quadratic complex which pass through any given point generate a quadric cone. At a *singular point* this cone has a double line and therefore breaks into two plane pencils. Correlatively, the rays which lie in any given plane envelope a conic which in the case of a *singular plane* has a double tangent and therefore degenerates into two points. Each of these points, being the vertex of a pencil of rays, is a singular point, and similarly each of the planes at a singular point is a singular plane and contains one other singular point. A *singular ray* is a ray of the complex characterised by the fact that all the tangent linear complexes are special, and is the double line of the complex cone at a singular point and also the double tangent of the complex curve in a singular plane.

* Jessop, *The Line Complex*, Chs. VI and XVII; Sturm, *Liniengeometrie*, III, 1.

Thus at a singular point P the complex cone consists of two planes π_1, π_2 intersecting in a singular ray x. The complex conics in all planes through x touch x at P except for one plane π in which the complex conic consists of two points P_1, P_2 on x. This plane π is the common tangent plane of all the complex cones whose vertices are points on x, except when the vertex is P, in which case the tangent plane through x is indeterminate.

A fundamental theorem states that the locus of singular points P is the same as the envelope of singular planes π and is a quartic surface. The proof of this is instantaneous when coordinates are used, and follows geometrically from another fundamental theorem that the four singular points on any line have the same cross ratio as the four singular planes through it.

We have then a *singular surface* Φ of the fourth order and fourth class; x is a tangent line at P and meets Φ in the remaining two singular points P_1, P_2 on it. The four tangent planes through x are π repeated and π_1, π_2. The complex rays in π form two pencils whose vertices are P_1 and P_2; no line of the tangent pencil (P, π) is a ray except the singular ray x, unless P_1 or P_2 coincides with P; then x meets Φ in three consecutive points and is therefore an *inflexional tangent.* In this case one of the planes π_1 or π_2 coincides with π, and *all the tangent lines at P are rays.*

Associated with the quadratic complex C there is one set of six linear complexes, mutually apolar. The polars of every ray of C with respect to these complexes belong to C and hence C may be termed *self-polar** with respect to them. By their means we are able to group together certain singular points and planes; for the singular surface, being determined by the rays of the complex, must be invariant under the transformations determined by the linear complexes: in other words, all the points and planes obtained from any one singular point or plane by the correlations of the six apolar complexes are also singular points and planes.

In future we shall denote the singular surface by Φ. Any point P of Φ gives rise to a 16_6 configuration inscribed in and circumscribed about Φ, and the corresponding tangent plane π gives rise to another. Each of these configurations is inscribed in the other. The quartic section by the plane π contains its six

* The term *apolar* might also be used, for if the equation of the quadratic complex is suitably modified by means of the identical quadratic relation among the line coordinates, the transvectant formed from it and any one of the linear complexes vanishes identically.

null-points P_1, P_2, P_3, P_4, P_5, P_6 as well as the point of contact P which is a double point of the section. The line PP_r belongs to the rth linear complex, and the null-plane of P in this complex is the tangent plane at P_r. Since this plane passes through P, PP_r is a *bitangent*, touching Φ at P and P_r. Hence the bitangents are rays of the fundamental complexes and form six congruences of the second order and class.

§ 31. OUTLINE OF THE ALGEBRAICAL THEORY.

The simplicity of the algebraical treatment depends on the simultaneous reduction of the fundamental relation satisfied by the coordinates of any line and of the given quadratic complex to canonical forms*. These are taken to be

$$x_1^2 + x_2^2 + x_3^2 + x_4^2 + x_5^2 + x_6^2 = 0 \quad \ldots\ldots\ldots\ldots\ldots(1),$$

$$k_1x_1^2 + k_2x_2^2 + k_3x_3^2 + k_4x_4^2 + k_5x_5^2 + k_6x_6^2 = 0 \quad \ldots\ldots(2).$$

Then the coordinate complexes $x_s = 0$ are the six fundamental complexes with respect to which (2) is self-polar; the line transformations associated with them are effected by changing the signs of the coordinates, and it is obvious that (2) is unaltered by this procedure.

At a singular ray the tangent linear complex is special: then k_sx_s must be the coordinates of a line and therefore

$$k_1^2x_1^2 + k_2^2x_2^2 + k_3^2x_3^2 + k_4^2x_4^2 + k_5^2x_5^2 + k_6^2x_6^2 = 0 \ldots\ldots(3).$$

When (1), (2), (3) are satisfied the lines (x) and (kx) determine a singular point and a singular plane of the complex; they also determine a plane pencil of lines

$$y_s = k_sx_s - \mu x_s$$

satisfying the equations

$$\Sigma (k_s - \mu)^{-1} y_s^2 = 0, \quad \Sigma (k_s - \mu)^{-2} y_s^2 = 0.$$

Hence (y) is a singular ray of the complex obtained from (2) by replacing k_s by $(k_s - \mu)^{-1}$, and the corresponding singular point and singular plane are determined by (y) and $(k - \mu)^{-1}y$ and are therefore the same as before.

We have thus found a singly infinite system of complexes

$$\Sigma (k_s - \mu)^{-1} x_s^2 = 0 \quad \ldots\ldots\ldots\ldots\ldots\ldots\ldots(4),$$

which have the same singular points and planes, and are therefore

* First adopted by Klein, in his *Inauguraldissertation* (Bonn, 1868); *Math. Annalen*, XXIII, 539.

termed *cosingular*. The original complex (2) is the member of this system which corresponds to $\mu = \infty$.

By differentiating (2) we obtain $\Sigma k_s x_s dx_s = 0$ which is the condition that the pencil (x, kx) may have an envelope. Hence the locus of singular points is the same as the envelope of singular planes and is the singular surface Φ. Every tangent line of Φ is expressible in the form $(kx) - \mu(x)$, (x) being a singular ray of (2), and is a singular ray of the cosingular complex whose coefficients are $(k_s - \mu)^{-1}$. Hence the plane pencils of tangent lines to Φ are *projectively* related to each other and to the cosingular family in such a way that corresponding lines are singular rays of the same complex.

To determine the order of Φ we consider the singular points on any line (y). Since these points are the same for all the members of the cosingular family, we may select one which contains (y) and suppose its coefficients to be k_s. At each singular point on (y) there is a plane pencil of rays $(\lambda x + \mu y)$ containing (y) and the singular ray. The conditions for the line (x) are

$$\Sigma x_s y_s = 0, \quad \Sigma k_s x_s y_s = 0, \quad \Sigma k_s x_s^2 = 0,$$

determining a ruled surface of degree four which consists of four pencils whose vertices are the singular points on (y) and whose planes are the singular planes through (y); hence Φ is of the fourth order and fourth class.

The four generators of this degenerate scroll which meet any line (z) belong to a regulus determined by the equations

$$\Sigma x_s y_s = 0, \quad \Sigma k_s x_s y_s = 0, \quad \Sigma x_s z_s = 0,$$

and therefore cut their two common transversals (y) and (z) projectively. Whence follows the important theorem that the cross ratio of the singular points on any line is equal to the cross ratio of the singular planes through it.

It will be observed that the proof of this theorem is independent of the existence of Φ, and it may be used to prove that the locus of singular points is the same as the envelope of singular planes. For, if two out of four elements coincide, their cross ratio vanishes, and conversely if the cross ratio vanishes, at least two elements coincide. Hence the locus and the envelope have the same tangent lines and therefore coincide. It is important to notice that when the cross ratio vanishes the coincidences among the points and planes need not completely correspond, and there may exist lines which do not bear reciprocal relations to the singular surface.

§ 32. ELLIPTIC COORDINATES.

When (x) is any line, $\Sigma x_s^2=0$ and the equation $\Sigma(k_s-\mu)^{-1}x_s^2=0$ gives four values of μ, namely the parameters of the cosingular complexes which contain (x). This equation is of great importance in determining the relation of (x) to Φ; its roots are called the *elliptic coordinates* of the line.

Let (x) cut Φ in P_1, P_2, P_3, P_4, and let the tangent planes through (x) touch Φ at A, B, C, D, respectively. The lines AP_1, AP_2, AP_3, AP_4 belong to the tangent pencil at A and are respectively singular rays of four cosingular complexes whose parameters are $\mu_1, \mu_2, \mu_3, \mu_4$, say. The rays of the first of these complexes which lie in the singular plane P_1AP_4 form two pencils one of which has its vertex at P_1 and therefore contains (x); hence (x) is a ray of each of the complexes and $\mu_1, \mu_2, \mu_3, \mu_4$ are its elliptic coordinates. We have seen that the lines AP_s of the tangent pencil at A are projectively related to the parameters μ_s and hence the same is true of the points P_s and the lines projecting them from B, C, D. Now four points can be projectively related to themselves in only four ways, and further AP_1 and BP_1 cannot be singular rays of the same complex if (x) is not a tangent to Φ at P_1; thus if we suppose that AP_1, BP_2, CP_3, DP_4 are corresponding tangent lines and singular rays of the same complex, the four sets of singular rays must be

AP_1, BP_2, CP_3, DP_4 in complex μ_1,

AP_2, BP_1, CP_4, DP_3 „ „ μ_2,

AP_3, BP_4, CP_1, DP_2 „ „ μ_3,

AP_4, BP_3, CP_2, DP_1 „ „ μ_4.

By a reciprocal course of reasoning the four singular planes cut the tangent plane at P_1 in singular rays of the complexes $\mu_1, \mu_2, \mu_3, \mu_4$; and for the tangent planes at P_2, P_3, P_4 the complexes are permuted without altering the cross ratios of their parameters. Hence the pencil of tangent planes through (x) is projectively related to the elliptic coordinates of (x) in any one of four equivalent orders.

Since $\mu_1, \mu_2, \mu_3, \mu_4$ are the roots of

$$\Sigma\,(k_s-\mu)^{-1}\,x_s^2=0,$$

we must have for all values of μ

$$\Sigma\,(\mu-k_s)^{-1}\,x_s^2\equiv C\,(\mu-\mu_1)\,(\mu-\mu_2)\,(\mu-\mu_3)\,(\mu-\mu_4)/f(\mu)$$

where

$$f(\mu) = (\mu - k_1)(\mu - k_2)(\mu - k_3)(\mu - k_4)(\mu - k_5)(\mu - k_6)$$

and

$$C = \Sigma k_s x_s^2.$$

On multiplying by $\mu - k_s$ and then putting $\mu = k_s$ we obtain the line coordinates in terms of the elliptic coordinates in the form

$$x_s^2 = C(k_s - \mu_1)(k_s - \mu_2)(k_s - \mu_3)(k_s - \mu_4)/f'(k_s).$$

§ 33. CONJUGATE SETS.

We see that there is not a one-one correspondence between the singular points on a line and the singular planes through it; but on the other hand, the three different partitions into two pairs correspond. Thus if we denote the four tangent planes by their points of contact A, B, C, D we see that the partitions of points

$$P_1P_4,\ P_2P_3 \qquad P_2P_4,\ P_3P_1 \qquad P_3P_4,\ P_1P_2$$

correspond to the partitions of planes

$$AD,\ BC \qquad BD,\ CA \qquad CD,\ AB,$$

and the elements of each set may be permuted provided the cross ratio remain unchanged.

Any pair of points, such as P_1P_4, and either of the corresponding pairs of planes, such as BC, are said to form a *conjugate set**. When any two points P_1P_4 are given a pair of planes forming with them a conjugate set may be constructed by selecting any one, for example μ_2, of the four complexes which contain P_1P_4; the cone of rays through P_1 breaks into two planes of which one, B, contains P_1P_4, and similarly the plane C forms part of the complex cone at P_4, and these two planes complete the set. By taking all four complexes in turn only two different pairs of planes are obtained; for example, the plane pencils (P_1, C) and (P_4, B) belong to the same complex μ_3.

Conversely, the planes and vertices of any two plane pencils belonging to the same complex and having a common ray form a conjugate set. For let P and P' be the vertices, which must be singular points, and let A, A' be the points of contact of the plane pencils at P, P' respectively with the singular surface, so that AP and $A'P'$ are singular rays of the same complex; it follows, by comparison with the preceding work, that AP' and $A'P$ must be singular rays of *another* of the complexes containing PP'.

* Klein, *Math. Annalen*, XXVII, 107.

The four points in which any line cuts Φ are determined by a biquadratic equation, and the four tangent planes through the same line by another equation. The relation between these equations is that the cross ratios of the roots are equal. This condition implies that the ratios of the roots of the reducing cubic are the same for both equations, and hence that if one biquadratic is expressed as the product of two quadratic factors, the other biquadratic can be similarly separated into two factors by rational means. Hence when three elements of a conjugate set are given the determination of the fourth depends upon finding the second root of a quadratic equation of which one root is given, and can therefore be effected by *rational* processes.

§ 34. KLEIN'S TETRAHEDRA.

With reference to a particular complex (λ) of the cosingular family let the singular ray of the tangent pencil at the point A cut Φ in A_1 and A_2 so that A_1, A_2, and A repeated are the four singular points on this ray. Similarly let BB_1B_2, CC_1C_2, DD_1D_2 be the singular rays at B, C, D respectively. The lines A_1P_s, A_2P_s, etc. are rays of the complex, and hence the eight points $A_1 \ldots D_2$ lie on the complex cones at each of P_1, P_2, P_3, P_4. Since P_s is a singular point the complex cone at P_s breaks up into two planes each of which contains four of the eight associated points.

If the parameter λ is taken equal to μ_1, the complex contains AP_1 as a singular ray; then BP_2, CP_3, DP_4 are also singular rays and we may suppose that A_2, B_2, C_2, D_2 coincide with P_1, P_2, P_3, P_4 respectively. One of the planes into which the complex cone at P_1 breaks up is $A_1P_2P_3P_4$ and so the other must be $P_1B_1C_1D_1$; similarly $A_1P_2C_1D_1$, $A_1B_1P_3D_1$, and $A_1B_1C_1P_4$ are planes of rays at P_2, P_3, P_4 respectively. Returning to the original complex we see that it is possible to name the eight associated points so that $A_1B_2C_2D_2$ and $A_2B_1C_1D_1$ are the planes of rays at P_1, and so on. We may express this by saying that the table

$$\begin{matrix} A_1 & B_1 & C_1 & D_1 \\ A_2 & B_2 & C_2 & D_2 \\ P_1 & P_2 & P_3 & P_4 \end{matrix}$$

is an incidence diagram, that is, the five points in a row and a column, but not in both, are coplanar. Each of the tetrahedra $A_1B_1C_1D_1$ and $A_2B_2C_2D_2$ is inscribed in the other, and the four lines joining corresponding vertices, and the four lines of intersection of corresponding faces, have a common transversal $P_1P_2P_3P_4$.

From what has been proved it follows that $P_1A_1P_2B_2$ is a twisted quadrilateral of rays of (λ) which are four edges of a tetrahedron, inscribed in Φ, whose faces touch Φ. Two adjacent sides of this quadrilateral determine one of the complex planes at their common point, so that the ends of any side and the two faces through that side form a conjugate set; for example, the pair of points P_1A_1 is conjugate to the pair of planes $P_1A_1P_2$, $P_1A_1B_2$. Now $P_1[A_1P_2]$ is a pencil of rays of (μ_1), so that by the property of conjugacy it follows that $A_1[P_1B_2]$ is a pencil of rays of the same complex (μ_1) and in particular A_1B_2 belongs to (μ_1). Hence $P_1A_1B_2P_2$ is a quadrilateral of rays of (μ_1) and similarly $P_2A_1B_2P_1$ is a quadrilateral of rays of (μ_2), and every edge of the tetrahedron is a common ray of two of the three complexes (λ), (μ_1), (μ_2).

Such a tetrahedron may be constructed by taking any three points on a tangent plane section and completing the three conjugate sets determined by this plane and pairs of points; the three new tangent planes meet on the surface. We see in this way that a given surface has ∞^5 inscribed and circumscribed tetrahedra*.

§ 35. RELATIONS OF LINES TO Φ.

We shall now consider the various equalities that can exist among the elliptic coordinates of a line, and its corresponding relations to Φ.

The elliptic coordinates are the roots of the equation in μ

$$\Sigma (k_s - \mu)^{-1} x_s^2 = 0$$

and are uniquely determined by the line (x); but conversely, an arbitrary set of roots determines thirty-two lines, polars of each other with respect to the fundamental complexes. In what follows, the line (x) means any one of these.

If two roots are equal to λ, then

$$\Sigma (k_s - \lambda)^{-1} x_s^2 = 0 \text{ and } \Sigma (k_s - \lambda)^{-2} x_s^2 = 0,$$

and (x) is a singular ray of the complex (λ) and is a tangent to Φ. Put $x_s = (k_s - \lambda) y_s$, then (y) is a singular ray of $\Sigma k_s x_s^2 = 0$. The

* "Ausgezeichnete Tetraeder," Klein, *Math. Annalen*, XXVII, 110.

elliptic coordinates of x are the roots of

$$\Sigma (k_s - \lambda)^2 (k_s - \mu)^{-1} y_s^2 = 0$$

or

$$(\mu - \lambda)^2 \Sigma (k_s - \mu)^{-1} y_s^2 = 0$$

or

$$(\mu - \lambda)^2 (\mu - \mu_1)(\mu - \mu_2) = 0$$

where μ_1 and μ_2 depend only on (y). As λ varies we get all the tangents of a plane pencil, μ_1 and μ_2 remaining constant; among these tangents are six special ones given by $\lambda = k_s$. Taking $\lambda = k_1$ we deduce from the preceding equations

$$x_1 = 0, \qquad \sum_2^6 (k_s - k_1)^{-1} x_s^2 = 0.$$

This shows that the line is a bitangent, for the null-plane of the point of contact and the null-point of the tangent plane in the linear complex $x_1 = 0$ are a plane and point of Φ by the invariant property of Φ, and all four elements are incident with (x). We infer that as μ_1 and μ_2 vary, the tangent lines for which $\lambda = k_1$ are bitangents; it is easy to prove that all the bitangents generate in this way six congruences of the second order and class.

Three roots can be equal only if $\lambda = \mu_1$ or μ_2. Now we have seen that μ_1 and μ_2 are the same for all the lines of a tangent pencil, and hence the whole pencil belongs to the cosingular complexes whose parameters are μ_1 and μ_2; but this can be the case only when the singular ray is an inflexional tangent (p. 52). Thus, when (x) is a singular ray of $\Sigma k_s x_s^2 = 0$ and μ_1, μ_2 are the roots of

$$\Sigma (k_s - \mu)^{-1} x_s^2 = 0$$

we have proved that $(kx) - \mu_1 (x)$ and $(kx) - \mu_2 (x)$ are the inflexional tangents at the point of contact, and the elliptic coordinates of these lines are $(\mu_1, \mu_1, \mu_1, \mu_2)$ and $(\mu_1, \mu_2, \mu_2, \mu_2)$ respectively.

If two pairs of roots are equal, say $\mu_1 = \mu_2$ and $\mu_3 = \mu_4$, then, omitting a factor of proportionality,

$$x_s = (k_s - \mu_1)(k_s - \mu_3) \{f'(k_s)\}^{-\frac{1}{2}}$$

and x_s, $(k_s - \mu_1)^{-1} x_s$, $(k_s - \mu_3)^{-1} x_s$ are the coordinates of three mutually intersecting lines. Every line which meets all three has line coordinates of the form

$$(ak_s^2 + bk_s + c) \{f'(k_s)\}^{-\frac{1}{2}}$$

and consequently has two pairs of equal elliptic coordinates and touches Φ. Hence if the three lines are concurrent their common point is a node of Φ and if they are coplanar their plane is a trope. There are thirty-two cases obtained by taking all possible combinations of signs for the radicals, showing that Φ possesses sixteen nodes and sixteen tropes. The identity of the singular surface with Kummer's surface is thus completely established. It is worth noticing that for a line through a node one pair of points coincide and two pairs of planes, and for a line in a trope two pairs of points coincide and one pair of planes: in both cases two pairs of elliptic coordinates are equal; but, on the other hand, in the case of a proper bitangent two pairs of points and two pairs of planes coincide while only two elliptic coordinates are equal.

If four roots are equal, then

$$x_s = (k_s - \mu)^2 \{f'(k_s)\}^{-\frac{1}{2}}.$$

Either (x) passes through a node and three points and four tangent planes coincide, so that (x) is a generator of the tangent cone at the node; or (x) lies in a trope and four points and three planes coincide, so that (x) touches the singular conic. Hence the tangent lines at a node and the tangents to a singular conic are projectively related to the tangents at an ordinary point, corresponding lines being singular rays of the same complex.

§ 36. ASYMPTOTIC CURVES.

We have seen that if (x) is a singular ray of $\Sigma k_s x_s^2 = 0$ the inflexional tangents of the pencil $(kx) - \mu(x)$ are given by the quadratic in μ

$$\Sigma (k_s - \mu)^{-1} x_s^2 = 0.$$

If the roots are μ_1 and μ_2 the elliptic coordinates of the inflexional tangents are $(\mu_1, \mu_1, \mu_1, \mu_2)$ and $(\mu_1, \mu_2, \mu_2, \mu_2)$. μ_1 and μ_2 may be regarded as parameters associated with a point on the surface Φ, and the equation of any curve on the surface may be expressed by a single relation between these parameters.

We shall now prove the remarkable theorem that the asymptotic curves are given by

$$\mu_1 = \text{const.} \quad \text{or} \quad \mu_2 = \text{const.}$$

An inflexional tangent at P can be regarded as a tangent to the surface at two consecutive points P, P'. It is a ray of the cosingular complex which has for a singular ray the other inflexional tangent at P. Hence the tangent pencil at P' contains a ray $P'P$ of this complex which is not singular; therefore one of the inflexional tangents at P' is a singular ray, and by continuity it must be the one which is not nearly coincident with $P'P$. Hence at different points of the same asymptotic curve the *other* inflexional tangents (not touching the curve) are singular rays of the *same* cosingular complex.

On account of the importance of this theorem we give an analytical proof.

If we want to find the value of $d\mu_1/d\mu_2$ along the curve whose tangent is (y), where $y_s = (k_s - \mu)\, x_s$, we must express the condition in elliptic coordinates that (y) may intersect its consecutive position. The condition is

$$\Sigma dy_s^2 = 0$$

or

$$\Sigma \,\{(k_s - \mu)\, dx_s - x_s\, d\mu\}^2 = 0$$

or

$$\Sigma\, (k_s - \mu)^2\, dx_s^2 = 0,$$

and on substituting $(k_s - \mu_1)(k_s - \mu_2)/f'(k_s)$ for x_s^2 this becomes

$$\Sigma \frac{(k_s - \mu)^2}{f'(k_s)} \left(\frac{k_s - \mu_2}{k_s - \mu_1} d\mu_1^2 + 2d\mu_1 d\mu_2 + \frac{k_s - \mu_1}{k_s - \mu_2} d\mu_2^2\right) = 0,$$

or

$$\frac{(\mu_1 - \mu)\, d\mu_1}{\sqrt{f(\mu_1)}} = \frac{(\mu_2 - \mu)\, d\mu_2}{\sqrt{f(\mu_2)}}.$$

Now along an asymptotic curve $\mu = \mu_1$ and then this differential equation can be integrated and gives

$$\mu_2 = \text{const.}$$

Similarly when $\mu = \mu_2$, $\mu_1 = \text{const.}$ Hence along every asymptotic curve the parameter of the other inflexional tangent is constant.

We have seen that every bitangent has two of its elliptic coordinates equal to one of the k_s; the remaining two are parameters of the inflexional tangents at *either* point of contact. Hence the points of an asymptotic curve can be joined in pairs by bitangents. As μ_2 varies the line

$$x_s = (k_s - k_i)(k_s - \mu_1)^{\frac{1}{2}}(k_s - \mu_2)^{\frac{1}{2}}\{f'(k_s)\}^{-\frac{1}{2}}$$

describes a scroll each generator of which touches Φ at two points on the asymptotic curve associated with μ_1.

The equation $\Sigma a_s x_s = 0$, when rationalised, is of degree 8 in μ_2, showing that the scroll is of degree 8. The complete intersection with the quartic surface consists of the asymptotic curve repeated, which is therefore of order 16.

Two asymptotic curves cut at points where one touches the scroll of bitangents circumscribing Φ along the other, that is at thirty-two points beside the nodes. These points are obtained from any one point by drawing successive bitangents and are derived from a point or its tangent plane by an even or odd number of correlations.

When $\mu_1 = \mu_2$, the inflexional tangents coincide. All the elliptic coordinates are equal and the line is either a generator of a nodal cone or a tangent of a singular conic. The cusp locus for asymptotic curves reduces to the sixteen nodes, and their envelope is the sixteen conics which form the *parabolic curve*. Each trope is a stationary osculating plane of the asymptotic curves so that the sixteen points of intersection with a trope are accounted for by two at each node and four at the point of contact. The parameter of this point, regarded as belonging to the singular conic, is the same as the parameter of the asymptotic curve.

§ 37. PRINCIPAL ASYMPTOTIC CURVES.

Among the asymptotic curves are six *principal asymptotic curves* corresponding to $\mu_1 = k_i$. The inflexional tangents of the surface which are not tangents of the curve have parameters k_i, k_i, k_i, μ and therefore belong to one of the six systems of bitangents, and hence have four-point contact. These curves pass once through the nodes and touch the singular conics there. The coordinates of a tangent having four-point contact are given by

$$x_s^2 = (k_s - k_i)^3 (k_s - \mu)/f'(k_s),$$

and the rationalised equation of intersection with a given line is of degree 8 in μ. Hence the four-point contact tangents corresponding to any principal asymptotic curve generate a scroll of degree 8 touching Φ all along an octavic curve.

The principal asymptotic curves occur as *repeated* curves in the family just as the fundamental linear complexes occur repeated in the family of cosingular complexes, and this accounts for the lowering of degree*.

On putting two elliptic coordinates equal to k_1 and k_2 respectively, the line coordinates x_1 and x_2 vanish, and so the various combinations of sign give only eight lines. Hence two principal asymptotic curves cut in eight points, besides the nodes, where they touch singular conics. At any common point both inflexional tangents have four-point contact, and the whole pencil of tangents belongs to the congruence (12). Hence the eight common points lie by fours on the directrices of (12), which are so related that the tangent planes through each touch at the points

* Compare the occurrence of a straight line among the projections of a given conic, and of a parabola among the harmonograms $x = \cos(t - a)$, $y = \cos 2t$.

on the other. Pairs of directrices are the only lines having this property*.

§ 38. THE CONGRUENCE OF SECOND ORDER AND CLASS.

We have seen that every bitangent of Φ belongs to one of six congruences, of which one is

$$x_1 = 0,\ (k_2 - k_1)^{-1} x_2^2 + (k_3 - k_1)^{-1} x_3^2 + (k_4 - k_1)^{-1} x_4^2 + (k_5 - k_1)^{-1} x_5^2 + (k_6 - k_1)^{-1} x_6^2 = 0.$$

Conversely, every ray of this congruence is a bitangent of Φ, for one elliptic coordinate μ_1 is equal to k_1 and the others must satisfy

$$\sum_{s=2}^{6} (k_s - \mu_2)(k_s - \mu_3)(k_s - \mu_4)/f'(k_s) = 0,$$

or

$$(k_1 - \mu_2)(k_1 - \mu_3)(k_1 - \mu_4) = 0,$$

so that a second elliptic coordinate must be equal to k_1.

Φ is therefore the *focal surface* of this congruence and we infer that the general Kummer surface is the focal surface of six congruences, which may be called *confocal*, and that the six linear complexes containing them are mutually apolar. We shall see presently that the six quadratic complexes, each of which is to a certain extent arbitrary, may be taken to be cosingular.

The preceding congruence is of a general character, for every (2, 2) congruence is contained in a linear complex, which may be taken for one of the coordinates in an apolar system unless the linear complex is special; then, in order to reduce the general congruence to the preceding form we have to reduce two quadratic forms in the remaining five coordinates simultaneously to sums of squares.

Thus the theory of the focal surface of a congruence is made to depend on the previously developed theory of the singular surface of a complex. In the following section is given a short independent account.

§ 39. SINGULARITIES OF THE CONGRUENCE.

The equations of the congruence being $x_1 = 0$, $\Sigma \lambda_s x_s^2 = 0$, through every point in space pass two distinct rays, the intersections of the *plane*

$$x_1 = 0,$$

and the *cone*

$$\lambda_2 x_2^2 + \lambda_3 x_3^2 + \lambda_4 x_4^2 + \lambda_5 x_5^2 + \lambda_6 x_6^2 = 0.$$

* For further particulars concerning asymptotic curves the following references may be consulted: Klein, *Math. Annalen*, v, 278, xxiii, 579; Reye, *Crelle*, xcvii, 242; Salmon-Fiedler, *Geometrie des Raumes*, ii, 491.

The focal surface is the locus of points through which the two rays coincide. We have to prove that it is of the fourth order and class and possesses sixteen nodes at the singular points of the congruence.

Coincidence arises from a special situation of this plane and cone and can occur only when

(1) the plane and cone touch,

(2) the cone has a double line and the plane passes through it,

(3) the cone breaks into the plane and another plane.

Case (1) occurs twice on each ray (x) of the congruence, for if (y) is a consecutive ray,

$$y_1 = 0, \qquad \Sigma\lambda_s x_s y_s = 0,$$

so that (x) and (y) are common rays of all the complexes whose coefficients are

$$\mu, \quad \lambda_2 x_2, \quad \lambda_3 x_3, \quad \lambda_4 x_4, \quad \lambda_5 x_5, \quad \lambda_6 x_6$$

for different values of μ. Now two rays of a linear congruence do not intersect except on the directrices; these are the lines whose coordinates are the preceding set of coefficients when

$$\mu^2 + \Sigma\lambda_s^2 x_s^2 = 0.$$

Let them be called (ξ) and (ξ'); then the two consecutive rays through the point (x, ξ) pass through consecutive points of (ξ') and therefore lie in the plane (x, ξ'); similarly the null-plane of the point (x, ξ') in the complex $x_1 = 0$ contains (ξ). Let $(x + dx)$ be any consecutive ray; then

$$dx_1 = 0, \qquad \Sigma\lambda_s x_s dx_s = 0,$$

whence $\qquad \Sigma\xi_s dx_s = 0, \qquad \Sigma\xi_s' dx_s = 0,$

which prove that the pencils (x, ξ), (x, ξ') have envelopes. These are the focal surface Φ, and we have seen that (x) touches it at two points and meets it at no others. Hence the surface is of the fourth order, and reciprocal reasoning shows that it is also of the fourth class.

Case (2) arises when (x) is a singular ray of the quadratic complex so that

$$\Sigma\lambda_s^2 x_s^2 = 0.$$

Let Λ be the singular surface of the complex

$$\lambda_2 x_2^2 + \lambda_3 x_3^2 + \lambda_4 x_4^2 + \lambda_5 x_5^2 + \lambda_6 x_6^2 = 0,$$

then the pencil $(x, \lambda x)$ of tangent lines of Λ belongs to the complex $x_1 = 0$, and the singular ray (x) counts as two intersections of the null-plane with the complex cone at the point $(x, \lambda x)$. Hence Λ and Φ *touch* at this point. Since $\Sigma\lambda_s^2 x_s^2 = 0$, the lines (ξ), (ξ') coincide and (x) has four-point contact with Φ.

In order that case (3) may arise, every line of the pencil $(x, \lambda x)$ must be a ray of the quadratic complex and hence (x) must be an inflexional tangent of Λ. These rays are given by the equations

$$x_1 = 0,\quad \Sigma x_s^2 = 0,\quad \Sigma\lambda_s x_s^2 = 0,\quad \Sigma\lambda_s^2 x_s^2 = 0,\quad \Sigma\lambda_s^3 x_s^2 = 0,$$

and are sixteen in number. The fact that every line of the pencil $(x, \lambda x)$ is a bitangent of Φ shows that their plane is a trope. These sixteen pencils form a 16_6 configuration of nodes and tropes of Φ, and the complete identity of Φ with a Kummer surface is established.

§ 40. RELATION BETWEEN Φ AND Λ.

We have seen that at every common point of Φ and Λ the two tangent planes coincide and the singular ray is a tangent having four-point contact with Φ. Hence the two surfaces touch along a principal asymptotic curve of Φ. Since the osculating plane of the curve coincides with the tangent plane of the surface, the curve must be also asymptotic on Λ: being an octavic, it is a principal asymptotic curve of Λ: and, counted twice, is the complete intersection of the two surfaces. Thus the relation between the surfaces is mutual; the nodes of each lie on the other and the tropes of each touch the other.

By comparing the equations

$$\Sigma\lambda_s x_s^2 = 0 \text{ and } \Sigma\,(k_s - k_1)^{-1}\,x_s^2 = 0,$$

we see that Φ is the singular surface of the quadratic complex

$$\lambda_2^{-1}x_2^2 + \lambda_3^{-1}x_3^2 + \lambda_4^{-1}x_4^2 + \lambda_5^{-1}x_5^2 + \lambda_6^{-1}x_6^2 = 0.$$

Since Λ is the singular surface of

$$\lambda_2 x_2^2 + \lambda_3 x_3^2 + \lambda_4 x_4^2 + \lambda_5 x_5^2 + \lambda_6 x_6^2 = 0,$$

we see again that the relation is mutual.

The addition of the same number to all the λ_s changes the first complex into another of the same cosingular family and so does not affect Φ. Hence for different values of λ_1, the singular surfaces of the complexes

$$\lambda_1 x_1^2 + \lambda_2 x_2^2 + \lambda_3 x_3^2 + \lambda_4 x_4^2 + \lambda_5 x_5^2 + \lambda_6 x_6^2 = 0$$

touch each other and Φ along the same curve, which passes through all their nodes.

§ 41. CONFOCAL CONGRUENCES.

Consider more generally the relation between the singular surface S of a quadratic complex $\Sigma k_s x_s^2 = 0$ and the focal surface F of the intersection of the same complex with a linear complex $z_1 = 0$ not self-polar with it.

Every singular tangent pencil of F forms part of a degenerate complex cone and hence has its vertex on S and its plane tangent to S at some other point. Hence the 16_6 configuration of nodes and tropes of F is inscribed in and circumscribed about S. Again the complex cone at a node of S consists of a repeated plane and so the node lies on F; reciprocally the tropes of S touch F.

At any common point P of S and F the null-plane of P must contain the singular ray, which is a ray of the congruence and therefore a bitangent of F, touching at P and Q. The null-plane of Q touches F at the other focal point P and the complex cone of Q along the singular ray QP. Hence it is the tangent plane to S at P ("π" of p. 52). Therefore S and F touch at all their common points and the curve of contact is an octavic passing through the nodes and touching the tropes of both surfaces.

Any trope of F meets this curve of contact in eight points lying on a conic; six of these are nodes $N_1, N_2, N_3, N_4, N_5, N_6$ of F and the remaining two coincide at a point O where the trope touches S. The null-point of this plane is a singular point of the congruence and a node on F, say N_1. All the lines through N_1 in the trope are rays of the congruence, and ON_1, being a tangent to S at O, is a singular ray of the complex. Now F is the singular surface of a complex for which $z_1 = 0$ is one of six fundamental linear complexes, and its equation may be taken to be

$$\lambda_1 z_1^2 + \lambda_2 z_2^2 + \lambda_3 z_3^2 + \lambda_4 z_4^2 + \lambda_5 z_5^2 + \lambda_6 z_6^2 = 0.$$

If we had started with the linear complex $z_2 = 0$ and that quadratic complex of the cosingular family $\Sigma (k_s - \lambda)^{-1} x_s^2 = 0$ which has ON_2 for a singular ray, N_2 being the null-point of the trope for $z_2 = 0$, the singular surface and the curve of contact would be the same as before, and the tangent plane at O determines the remaining fifteen planes of a circumscribed 16_6 configuration which are the tropes of F, so that the focal surface would be unaltered.

Hence when a congruence is given as the intersection of a quadratic and a linear complex, it is in general possible to find five other cosingular quadratic complexes and five other mutually apolar linear complexes so that, taken in pairs, they form, in all, six *confocal congruences*.

The parameters of the cosingular complexes are projectively related to the positions of the nodes on a conic of F, that is, to the coefficients λ_s, so that F, the singular surface of $\Sigma\lambda_s z_s^2 = 0$, is the focal surface of the six congruences

$$z_i = 0, \quad \Sigma_s (k_s - \lambda_i)^{-1} x_s^2 = 0, \qquad (i = 1, \dots 6)$$

and by symmetry S, the singular surface of $\Sigma k_s x_s^2 = 0$, is the focal surface of the six congruences

$$x_i = 0, \quad \Sigma_s (\lambda_s - k_i)^{-1} z_s^2 = 0,$$

and it is easy by comparing these with the equations

$$x_i = 0, \quad \Sigma (k_s - k_i)^{-1} x_s^2 = 0,$$

to deduce that the two sets of coordinates x and z are connected by the orthogonal transformation

$$z_r = \Sigma_s a_{rs} x_s,$$

where $\quad a_{rs}^2 (k_r - \lambda_s)^2 f'(k_r) \phi'(\lambda_s) = -f(\lambda_s) \phi(k_r)$

$$f(\theta) \equiv (\theta - k_1)(\theta - k_2)(\theta - k_3)(\theta - k_4)(\theta - k_5)(\theta - k_6),$$

$$\phi(\theta) \equiv (\theta - \lambda_1)(\theta - \lambda_2)(\theta - \lambda_3)(\theta - \lambda_4)(\theta - \lambda_5)(\theta - \lambda_6).$$

We can now see how when F is given S can be constructed geometrically in ∞^6 ways. At any point P of F draw the tangent plane and in it draw any conic through P cutting the section again in six points. A partition of these into two triangles determines the bases of two of Klein's tetrahedra having a common vertex; in this way ten more points and fifteen more planes are found which complete the configuration of nodes and tropes of S.

For different values of a_1 the singular surfaces of the complexes

$$\Sigma k_s x_s^2 + x_1 \Sigma a_s x_s = 0$$

touch the focal surface of the congruence given by this equation and $x_1 = 0$ along the same octavic curve, and for *ten* values of a_1 the discriminating sextic has a pair of equal roots, and the corresponding singular surface has a pair of coincident nodes in each trope and therefore a *nodal line*.

CHAPTER VI.

PLÜCKER'S COMPLEX SURFACE.

§ 42. TETRAHEDRAL COMPLEXES.

Plücker devotes the greater part of his *Neue Geometrie des Raumes* to the study of the surface named after him, with the intention of making clearer the arrangement of rays in a quadratic complex. The surface is the focal surface of a *special* congruence contained in the complex, and is therefore the locus of complex conics in planes through a line and at the same time the envelope of cones with vertices on that line. It is a degenerate form of Kummer surface due to two of the linear complexes with respect to which the quadratic complex is self-polar becoming special and coinciding, and is the singular surface of a quadratic complex with six *double lines**.

We have seen that every congruence of the second order and class is contained in one linear complex and many quadratic complexes. It is natural at the beginning of the investigation to choose the simplest complex, and we have the theorem that *every general quadratic congruence is contained in forty tetrahedral complexes.* To prove this we recall that the singular and focal surfaces touch along an octavic curve passing through their thirty-two nodes; if then the singular surface is four planes, the curve of contact must be four conics intersecting in four nodes. These sets of planes form "Rosenhain tetrahedra," of which there are eighty†. Now the linear complex containing the congruence is one of the fundamental complexes for the focal surface and we must reject those Rosenhain tetrahedra of which four edges are rays of this fundamental complex, for in that case the focal surface is easily shown to be a repeated quadric. Now any Rosenhain tetrahedron, such as

* Sturm, III, 355.

† See Chap. VII. (p. 78).

0, 23, 31, 12, has its edges belonging by fours to three fundamental complexes 1, 2, 3 so that by rejecting those whose edges belong to a particular one the number of available tetrahedra is reduced to forty.

We therefore start with a tetrahedral complex and select the rays cutting a given line. We shall take the singular surface of the complex for tetrahedron of reference and use current point coordinates x_1, x_2, x_3, x_4 and plane coordinates u_1, u_2, u_3, u_4 and line coordinates

$$p_{12} = q_{34} = x_1 y_2 - x_2 y_1 = u_3 v_4 - u_4 v_3$$

etc. The following abbreviations are useful:

$$\xi_i = \Sigma_s q_{is} x_s, \quad v_i = \Sigma_s p_{is} u_s,$$

so that $\xi_i = 0$ are the planes through a line and the corners of reference, and $v_i = 0$ are the points where the line cuts the faces of reference; they are, in this case, the singular planes through a line and the singular points on it. Again ξ_i are the coordinates of the plane through (p) and (x), and v_i are the coordinates of the point where (p) cuts (u).

§ 43. EQUATIONS OF THE COMPLEX AND THE COMPLEX SURFACE.

Let the equation of the complex be

$$a p_{14} p_{23} + b p_{24} p_{31} + c p_{34} p_{12} = 0,$$

which on account of the relation

$$p_{14} p_{23} + p_{24} p_{31} + p_{34} p_{12} = 0$$

gives $$p_{14} p_{23}/\alpha = p_{24} p_{31}/\beta = p_{34} p_{12}/\gamma,$$

where $$\alpha = b - c, \quad \beta = c - a, \quad \gamma = a - b,$$

and therefore $$\alpha + \beta + \gamma = 0.$$

The complex may be represented by an equation in mixed point and plane coordinates, as Plücker shows (p. 164). For the rays (p) in any plane (u) satisfy

$$u_1 p_{14} + u_2 p_{24} + u_3 p_{34} = 0,$$

whence $$\alpha u_1/p_{23} + \beta u_2/p_{31} + \gamma u_3/p_{12} = 0,$$

showing that the plane

$$\xi_4 \equiv p_{23} x_1 + p_{31} x_2 + p_{12} x_3 = 0$$

touches the cone

$$\sqrt{\alpha u_1 x_1} + \sqrt{\beta u_2 x_2} + \sqrt{\gamma u_3 x_3} = 0,$$

and therefore the rays in the plane

$$u_1 x_1 + u_2 x_2 + u_3 x_3 + u_4 x_4 = 0$$

touch the section of this cone. The last two equations give the complex curve enveloped by the rays in any plane (u). If on the other hand we regard (x) as a fixed point and the u_s as current plane coordinates the equations give the complex cone of rays through (x), as may be proved by exactly correlative reasoning. Hence the two equations completely represent the complex; the former can be replaced by the alternative forms

$$\sqrt{\gamma u_2 x_2} + \sqrt{\beta u_3 x_3} + \sqrt{\alpha u_4 x_4} = 0,$$
$$\sqrt{\gamma u_1 x_1} + \sqrt{\alpha u_3 x_3} + \sqrt{\beta u_4 x_4} = 0,$$
$$\sqrt{\beta u_1 x_1} + \sqrt{\alpha u_2 x_2} + \sqrt{\gamma u_4 x_4} = 0,$$

(in which the signs of the radicals are ambiguous), so that the want of symmetry is only apparent.

To find the equation of the complex surface in point coordinates we require the locus of the conic section

$$\sqrt{\alpha u_1 x_1} + \sqrt{\beta u_2 x_2} + \sqrt{\gamma u_3 x_3} = 0,$$
$$u_1 x_1 + u_2 x_2 + u_3 x_3 + u_4 x_4 = 0,$$

as the plane turns round a given line. As we shall not have further need for *current* line coordinates we shall take this line to be (p) or (q) and then the plane through it and any point (x') is

$$\xi_1' x_1 + \xi_2' x_2 + \xi_3' x_3 + \xi_4' x_4 = 0,$$

where $\xi_i' = \Sigma_s q_{is} x_s'$. Hence if (x') is any point on the conic section,

$$\xi_1'/u_1 = \xi_2'/u_2 = \xi_3'/u_3 = \xi_4'/u_4,$$

and the locus of the conic is

$$\sqrt{\alpha \xi_1 x_1} + \sqrt{\beta \xi_2 x_2} + \sqrt{\gamma \xi_3 x_3} = 0,$$

an equation which is equivalent to three others of similar form in virtue of the identities $\alpha + \beta + \gamma = 0$ and $\Sigma \xi_s x_s = 0$.

Next, to find the equation in plane coordinates we require the envelope of the *cone*

$$\sqrt{\alpha x_1 u_1} + \sqrt{\beta x_2 u_2} + \sqrt{\gamma x_3 u_3} = 0, \quad \Sigma x_s u_s = 0,$$

as the point (x) moves along the line (p). Since the coordinates of the point where (p) cuts (u) are v_i, the equation required is obtained by replacing x_i by v_i, and accordingly is

$$\sqrt{\alpha v_1 u_1} + \sqrt{\beta v_2 u_2} + \sqrt{\gamma v_3 u_3} = 0,$$

or three other equivalent forms.

From these equations it is evident that (p) is a nodal line, that $x_i = 0$, $\xi_i = 0$ are singular planes and $u_i = 0$, $v_i = 0$ are singular points of the surface.

§ 44. SINGULARITIES OF THE SURFACE.

We must next examine the sections through (p), the nodal line. Each consists of (p) counted twice and a complex conic

$$\sqrt{\alpha u_1 x_1} + \sqrt{\beta u_2 x_2} + \sqrt{\gamma u_3 x_3} = 0.$$

Regarded as a point locus this conic degenerates into a repeated line for four positions of the plane (u), namely those passing through the corners of reference. Thus, putting $u_1 = 0$, $u_2 = q_{12}$, $u_3 = q_{13}$, $u_4 = q_{14}$, we see that the plane

$$\xi_1 \equiv q_{12} x_2 + q_{13} x_3 + q_{14} x_4 = 0$$

touches the surface all along the line in which it meets the plane

$$\beta q_{12} x_2 - \gamma q_{13} x_3 = 0.$$

This line, which is called a *torsal line**, lies entirely on the surface; the singularity is of a tangential nature and consists in the fact that the tangent plane does not change as the point of contact moves along the line, as in the case of a generator of a torse or developable. The plane $\xi_1 = 0$, which is called a *pinch plane*, is a trope in which the conic of contact has broken into two lines, the nodal and torsal lines.

Regarded as an envelope the degenerate conic is touched by planes (u) satisfying

$$u_1 (\beta q_{12} u_3 + \gamma q_{13} u_2) = 0,$$

and consists of the two points $(1, 0, 0, 0)$ and $(0, \gamma/q_{12}, \beta/q_{13}, \alpha/q_{14})$ which are both nodes. These points are the vertices of pencils of rays of the complex in the singular plane $\xi_1 = 0$ and correspond to the points $A_1 A_2$ of § 34.

Thus in addition to the point singularities on the nodal line we have eight nodes lying by pairs on the four torsal lines; they are

$$\begin{array}{ll} A_1\,(1,\ 0,\ 0,\ 0) & A_2\,(\ 0,\ \ \gamma/q_{12},\ \beta/q_{13},\ \alpha/q_{14}) \\ B_1\,(0,\ 1,\ 0,\ 0) & B_2\,(\gamma/q_{21},\ \ 0,\ \ \alpha/q_{23},\ \beta/q_{24}) \\ C_1\,(0,\ 0,\ 1,\ 0) & C_2\,(\beta/q_{31},\ \alpha/q_{32},\ \ 0,\ \ \gamma/q_{34}) \\ D_1\,(0,\ 0,\ 0,\ 1) & D_2\,(\alpha/q_{41},\ \beta/q_{42},\ \gamma/q_{43},\ \ 0\ ). \end{array}$$

Correlatively there are four points on (p) for which the complex cone degenerates. Regarded as an envelope, the cone with vertex (x)

$$\sqrt{\alpha x_1 u_1} + \sqrt{\beta x_2 u_2} + \sqrt{\gamma x_3 u_3} = 0$$

* Plücker names this a *singular line*. See Sturm, II, 201; Cayley, VI, 334.

is a repeated line when $x_1 = 0$; then $x_2 = p_{12}$, $x_3 = p_{13}$, $x_4 = p_{14}$ and the repeated line has plane equations

$$\beta p_{12} u_2 - \gamma p_{13} u_3 = 0,$$
$$v_1 \equiv p_{12} u_2 + p_{13} u_3 + p_{14} u_4 = 0.$$

This line, which is called a *cuspidal axis**, is a singularity of exactly reciprocal character to that of a torsal line; every plane through it is a tangent plane to the surface at the same point $v_1 = 0$, and every plane through v_1 cuts the surface in a section having a cusp there. Hence $v_1, \ldots v_4$ may be called *cuspidal points.* They correspond to the points P_1, P_2, P_3, P_4 of p. 55. They are also called *pinch points*, because the two sheets of the surface touch each other there.

Regarded as a line locus the cone at v_1 breaks into a pair of planes joining the lines

$$x_1 (\beta p_{12} x_3 + \gamma p_{13} x_2) = 0, \quad x_4 = 0$$

to the point v_1 or $(0, p_{12}, p_{13}, p_{14})$; they are both tropes and intersect in the cuspidal axis which has point equations

$$x_1 = 0, \quad \gamma x_2/p_{12} + \beta x_3/p_{13} + \alpha x_4/p_{14} = 0.$$

In addition therefore to the four torsal planes through (p) there are eight tropes intersecting by pairs in the four cuspidal axes; their coordinates are obtained from those of the nodes $A_1 \ldots D_2$ by changing p_{rs} into q_{rs}. If we use the same letters to denote the points and planes, all the incidences can be exhibited at once in the table

$$\begin{matrix} A_1 & B_1 & C_1 & D_1 \\ A_2 & B_2 & C_2 & D_2 \end{matrix}$$

in which a row and a column, excluding their common member, contain the names of four coplanar points and four concurrent planes. This is the configuration of mutually inscribed tetrahedra which has already been described in § 34†.

§ 45. THE POLAR LINE.

The locus of the poles of the nodal line (p) with respect to complex curves in planes through (p) is a straight line, for it must lie on the polar plane of (p) with respect to every complex cone whose vertex is on (p). This *polar line*‡ cuts each torsal line and

* Plücker names this line a *singular axis*. See Cayley, VI, 123; Sturm, *Liniengeometrie*, III, 5; *Math. Ann.* IV, 249; Zeuthen, *Math. Ann.* IV, 1.

† Sturm, III, 1; Klein, *Math. Ann.* VII, 208.

‡ Weiler names this the *adjoint line*, *Math. Ann.* VII, 170.

is harmonically conjugate to (p) with respect to the two nodes on it; reciprocally, both lines cut each cuspidal axis and determine with it planes harmonically conjugate with respect to the two tropes through it.

The complex surface is the singular surface of a quadratic complex for which two fundamental complexes are special and coincide; for considering the situation of nodes on the eight tropes we see that two coincide at the intersection with the nodal line, which is the directrix of the special complex. It is easy to prove that the other four fundamental complexes are, using P_{rs} for current coordinates,

$$\alpha (P_{14}/p_{14} + P_{23}/p_{23}) + \beta (P_{24}/p_{24} + P_{31}/p_{31}) + \gamma (P_{34}/p_{34} + P_{12}/p_{12}) = 0,$$
$$P_{14}/p_{14} = P_{23}/p_{23}, \quad P_{24}/p_{24} = P_{31}/p_{31}, \quad P_{34}/p_{34} = P_{12}/p_{12}.$$

When the nodal line is at infinity the polar line becomes the locus of centres of parallel conic sections. Models of the surface in this case are described by Cayley, *Collected Papers*, VII, 298.

The singular surface of the quadratic complex

$$\frac{k_2-k_6}{k_2-k_1}x_2^2+\frac{k_3-k_6}{k_3-k_1}x_3^2+\frac{k_4-k_6}{k_4-k_1}x_4^2+\frac{k_5-k_6}{k_5-k_1}x_5^2=0$$

is the tetrahedron 0, 24, 46, 62.

The surface is unicursal, and the coordinates are expressible in terms of parameters λ, μ as follows

$$x_1=\frac{(b-c)(a+\lambda)^2}{u_1+\mu v_1}, \quad x_2=\frac{(c-a)(b+\lambda)^2}{u_2+\mu v_2},$$
$$x_3=\frac{(a-b)(c+\lambda)^2}{u_3+\mu v_3}, \quad x_4=\frac{(b-c)(c-a)(a-b)}{u_4+\mu v_4}.$$

The nodal line, polar line and three nodes, no two of which lie on the same torsal line, determine the remaining singularities.

§ 46. SHAPE OF THE SURFACE.

A special form of the configuration of two mutually inscribed tetrahedra consists of the corners of two rectangles placed in planes perpendicular to the line joining their centres, the sides of one being parallel to the sides of the other. These are the nodes of a Plücker surface possessing two planes of symmetry but otherwise exhibiting the features of the general case. The diagonals of the rectangles are the torsal lines and in this case intersect by pairs on the nodal line, which is here an axis of symmetry. In fig. 7 four of the conics through the nodes are ellipses, touching each other by pairs at the extreme pinch points on the nodal line; the other four are hyperbolas, touching each other by pairs at the other two pinch points.

Now since the pinch points are points on the nodal line at which the two tangent planes coincide they divide it into segments through which real and imaginary sheets of the surface pass alternately. Since the torsal lines are real both of the segments with real sheets are here finite. Such a segment is the common edge of two finite wedge-shaped pieces of the surface, the angle of the

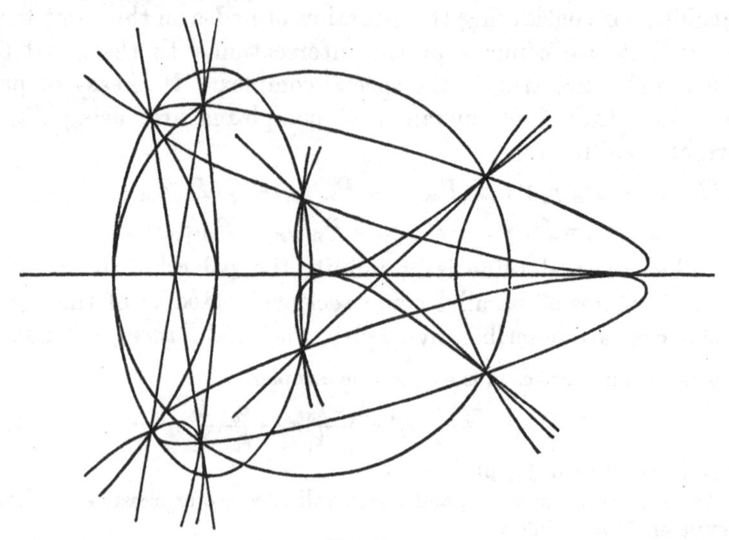

FIG. 7.

wedge varying from zero at the pinch points to the acute angle between the torsal lines. Each wedge contains two nodes, and the section of a pair of wedges by a plane through the nodal line is an ellipse. As this plane turns round the nodal line the elliptic section narrows until it becomes indefinitely thin and coincides with a finite portion of a torsal line terminated by the two nodes on it; as the plane continues to turn, the section immediately becomes a thin hyperbola, terminated by the same two nodes, which widens and remains hyperbolic until another torsal line is reached. Thus two nodes which are joined by *two* arcs of ellipses belong to an infinite piece of the surface, and there are four such pieces.

CHAPTER VII.

SETS OF NODES.

§ 47. GROUP-SETS.

On account of their importance in subsequent applications, various sets of points and planes of Kummer's configuration must be studied in detail. The terms *node* and *trope* will be used for convenience, and indicate the relation of the elements to the Kummer surface determined by them.

The 16_6 configuration is transformed into itself by fifteen collineations which together with identity form the group of sixteen members upon which the whole theory depends. These collineations have been expressed algebraically as simple linear transformations of point coordinates: geometrically each is effected by means of two opposite edges of a fundamental tetrahedron, any point being transformed into its harmonic conjugate with respect to the directrices of a fundamental congruence. To each collineation of the group corresponds a node and a trope represented by the same two-letter symbol, and it will appear that those sets of elements are most important which correspond to subgroups. Such a set is invariant for the subgroup and is changed by the other collineations into other sets each of which is invariant for the same subgroup. This set of different sets is here called a *group-set* and contains the whole configuration. The sets of a group-set are equivalent in the sense that they have the same projective relations to the configuration.

The incidence diagram (§ 5) is of great use in representing these sets of elements and in facilitating their enumeration, and is accordingly preferred to symbols. The effect of any collineation upon the diagram is simply to interchange two rows and at the same time the other two rows, or to interchange two columns and also the other two columns, or to make these changes simultaneously. If the names of the collineations are written in the diagram, the symbol which after these changes is in the first row and column is the name of the corresponding collineation.

§ 48. COMPARISON OF NOTATIONS.

For the general symmetrical treatment of the configuration in relation to the group of collineations the two-letter symbols are the most convenient, and agree with the notation subsequently used for thetafunctions. If preferred, Humbert's algorithm* may be used: it has the advantage of showing more clearly the position of each symbol in the incidence diagram. The two tables are given here for comparison:

11	12	13	14	*dd*	*bc*	*ca*	*ab*
21	22	23	24	*cb*	*aa*	*dc*	*bd*
31	32	33	34	*ac*	*cd*	*bb*	*da*
41	42	43	44	*ba*	*db*	*ad*	*cc*

It is, however, often convenient to isolate a particular element which is named 0, and the remaining fifteen are 12, 13, ... 56. Finally, it may be necessary to distinguish the six elements which are incident with a particular element from the remaining ten; the former are named 1, 2, 3, 4, 5, 6 and the latter by the partitions of these figures into two sets of three: of a pair of sets it is sufficient to name only one. These two nomenclatures are based upon the construction of the configuration from six apolar complexes and may be appropriately exhibited in two complementary diagrams (p. 45)

0	46	62	24	135	2	4	6
35	12	14	16	1	146	162	124
51	32	34	36	3	346	362	324
13	52	54	56	5	546	562	524

of which one represents nodes and the other tropes, but there is no essential geometrical correspondence between the elements in corresponding situations.

In these two systems the figures 1, 2, 3, 4, 5, 6 represent permutable operations which, repeated, produce identity, denoted here by 0; further 123456 = 0, so that every compound operation can be

* *Liouville*, sér. 4, IX, 55.

reduced to one containing three or fewer figures. By the *product* of symbols is meant the symbol of the product of operations.

A set of elements, expressed in the last notation, is called *odd* or *even* according to the number of single figure symbols, that is, the number of elements incident with a given element. If the parity is independent of a particular given element, it is an important feature of the set and has an essential geometrical significance.

§ 49. PAIRS AND OCTADS.

There are fifteen subgroups of two members each. Any one of these corresponds to a pair of nodes of which one is (*dd*) and gives rise to a group-set of eight pairs, of which examples are given in the diagrams

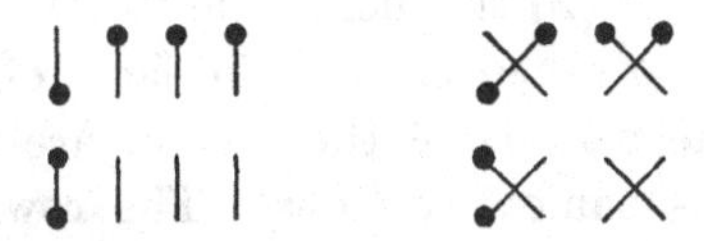

One pair of nodes possesses no special features in relation to the configuration as distinguished from another pair. Two nodes lie in two tropes, and are joined by one of the 120 *Kummer lines*. The pair is invariant under a subgroup of two members, namely identity and the collineation represented by the product of their symbols. The diagrams show that of the eight pairs of a group-set four are odd and four even, and, further, the partition into two sets of four pairs is invariant under the group. The four odd pairs form an odd *octad* and the four even pairs form an even *octad*; these two octads together make up a group-set, and are said to be *associated*. Hence there are fifteen couples of associated octads, one in each couple being odd and the other even.

An octad is represented in the diagram by two rows, or two columns, or two complementary rectangles, and thus corresponds to a bilinear identity among the sixteen linear forms (p. 31). It is noteworthy that the *terms* of the identity indicate the *pairs* of the octad. An octad of eight nodes is a group of *eight associated points* lying on four pairs of planes forming an octad of tropes.

The eight Kummer lines of a group-set cut a pair of directrices and determine on each three involutions whose double points are the corners of the three fundamental tetrahedra having that directrix for an edge.

The eight nodes of an octad can be joined by four Kummer lines in seven ways, and in six of these ways the lines belong to a regulus.

Two pairs from a group-set form a *tetrad*; if they are taken from different octads the tetrad is *odd*, and named after Rosenhain: if the pairs are taken from the same octad the tetrad is *even* and named after Göpel. In both cases the product of the four symbols is identity. These properties are sufficient to define the two kinds of tetrad and will be found to agree with the geometrical definitions given in the next two sections.

§ 50. EIGHTY ROSENHAIN ODD TETRADS.

These are tetrahedra whose corners are nodes and whose faces are tropes. Two of the nodes can be chosen arbitrarily, the third must lie in one of the two tropes containing them both, and then the fourth is determined. There are two types of diagram, in (1) the points lie in a line and represent the corners and faces of the same tetrahedron; in (2) the points lie in two lines and represent the corners of one tetrahedron and the faces of another; the faces of the first and the corners of the second are represented by the other points in the same two lines. The rows and columns give

(1)	● · · · ● · · · ● · · · ● · · ·	(2)	● · · · ● · · · · ● · · · ● · ·

eight tetrads of type (1) and each of the six pairs of rows and six pairs of columns give six tetrads of type (2), making *eighty* in all.

In symbols, type (1) is represented by

$$135,\ 1,\ 3,\ 5 \quad \text{or by} \quad 0,\ 35,\ 51,\ 13$$

and this shows how a Rosenhain tetrad is constructed. The nodes in any one trope are partitioned into two triangles 1, 3, 5 and 2, 4, 6: the other tropes through the sides of these triangles all pass through the point 135 . 246 which is the common corner of two tetrahedra having a common face. We infer that there are sixteen sets of ten tetrahedra having a common face.

All the tetrahedra of type (1) are self-polar with respect to one of the fundamental quadrics (p. 31). Since there are ten quadrics playing symmetrical parts in the configuration, we obtain in this way ten sets of eight tetrahedra, making, once more, eighty in all.

The Rosenhain tetrads are sufficiently characterised by the properties of being odd and invariant for at least one collineation. From the latter property it follows that the product of the symbols of a tetrad is identity and thence that the collineation which interchanges any two corners interchanges the other two at the same time. Hence each tetrad is invariant for a subgroup of four members and there are four tetrads in a group-set. Each set of eight which are self-polar with respect to the same quadric contains two group-sets: for example the four rows represent the tetrads of one group-set and the four columns those of another.

The faces of a Rosenhain tetrahedron contain all sixteen nodes, from which it follows that the four singular conics in the faces do *not* lie on a quadric surface.

§ 51. SIXTY GÖPEL EVEN TETRADS.

These are tetrahedra of nodes whose faces are not tropes, or tetrahedra of tropes whose corners are not nodes. There are two types of diagram according as the four points (1) form a rectangle or (2) lie on different rows and columns.

(1) (2)

Each tetrad can be divided in three ways into two pairs belonging to the same octad, and each of the thirty octads contains six tetrads, giving *sixty* tetrads in all.

The Göpel tetrads are sufficiently characterised by the properties of being even and invariant for at least one collineation. As in the case of odd tetrads, each is invariant for a subgroup and there are four tetrads in a group-set. Examples are

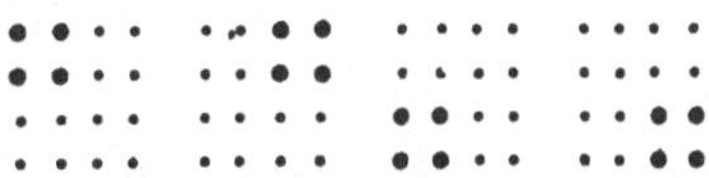

and

A typical representation in two-figure symbols is 0, 12, 34, 56 and we associate a group-set with a partition of six figures into three pairs. In order to construct a Göpel tetrahedron having a given trope for one face, we join the nodes in that trope in pairs by three lines; the other tropes through these lines complete the tetrahedron. Hence fifteen tetrads have one element common.

The tetrahedron 0, 12, 34, 56 of either nodes or tropes and the fundamental tetrahedron (12, 34, 56) belong to a desmic system. By taking the latter for reference it is easily seen that the third member of the system together with those obtained in a similar way from the other tetrahedra of the same group-set form a 16_6 configuration. Further the faces of a group-set of Göpel tetrahedra of nodes form another 16_6 configuration.

The four singular conics in a Göpel tetrahedron of tropes lie on a quadric.

§ 52. ODD AND EVEN HEXADS.

A set of six symbols whose product is identity is necessarily derived, in many ways, from two tetrads having a common element, after excluding that element. Hence an odd hexad of this kind is derived from an odd and an even tetrad, and is found to be a set of six elements from which the whole configuration can be linearly constructed (§ 6), and is named after Weber.

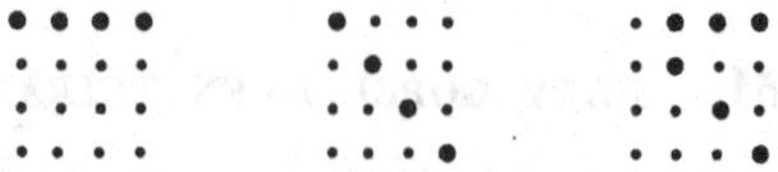

odd tetrad even tetrad odd hexad

A Weber hexad is not invariant for any collineation and hence a group-set contains sixteen hexads. The total number of hexads is 192.

The only other hexads of special interest are the Rosenhain hexads of coplanar nodes, or concurrent tropes. Every trope contains either two or six nodes of such a hexad, which is therefore even. The product of the symbols is identity, but this property is possessed also by other sets of six points. A Rosenhain hexad is not invariant for any collineation, and there is only one group-set.

CHAPTER VIII.

EQUATIONS OF KUMMER'S SURFACE.

§ 53. THE EQUATION REFERRED TO A FUNDAMENTAL TETRAHEDRON.

Taking a fundamental tetrahedron for reference* we write down the most general quartic equation which is unchanged by the operations of the group of § 4, that is by changing the signs of two coordinates or by the permutations of $(xyzt)$ into $(yxtz)$, $(ztxy)$ or $(tzyx)$. All the terms which are derived from any one term by these operations must have the same coefficient, and so the equation must have the form

$$x^4+y^4+z^4+t^4+2Dxyzt$$
$$+A\,(x^2t^2+y^2z^2)+B\,(y^2t^2+z^2x^2)+C\,(z^2t^2+x^2y^2)=0.$$

Now make the point $(\alpha, \beta, \gamma, \delta)$ a node: this gives four conditions which determine A, B, C, D in the forms

$$A=\frac{\beta^4+\gamma^4-\alpha^4-\delta^4}{\alpha^2\delta^2-\beta^2\gamma^2},\quad B=\frac{\gamma^4+\alpha^4-\beta^4-\delta^4}{\beta^2\delta^2-\gamma^2\alpha^2},\quad C=\frac{\alpha^4+\beta^4-\gamma^4-\delta^4}{\gamma^2\delta^2-\alpha^2\beta^2},$$

$$D=\frac{\alpha\beta\gamma\delta(\delta^2+\alpha^2-\beta^2-\gamma^2)(\delta^2+\beta^2-\gamma^2-\alpha^2)(\delta^2+\gamma^2-\alpha^2-\beta^2)(\alpha^2+\beta^2+\gamma^2+\delta^2)}{(\alpha^2\delta^2-\beta^2\gamma^2)(\beta^2\delta^2-\gamma^2\alpha^2)(\gamma^2\delta^2-\alpha^2\beta^2)},$$

and on eliminating $\alpha, \beta, \gamma, \delta$ there results the single condition among the coefficients

$$4-A^2-B^2-C^2+ABC+D^2=0.$$

Making use of the fundamental quadrics, we can write

$$2-A=\frac{(\delta^2+\alpha^2-\beta^2-\gamma^2)(\delta^2+\alpha^2+\beta^2+\gamma^2)}{(\alpha\delta-\beta\gamma)(\alpha\delta+\beta\gamma)}=4\frac{[da]\,[dd]}{[aa]\,[ad]},\text{ etc.}$$

$$D=\frac{64\alpha\beta\gamma\delta\,[da]\,[db]\,[dc]\,[dd]}{[aa]\,[ad]\,[bb]\,[bd]\,[cc]\,[cd]}=\frac{\alpha\beta\gamma\delta\,(2-A)\,(2-B)\,(2-C)}{[dd]^2}.$$

* The equation is worked out from an irrational form by Cayley, *Coll. Papers*, x, 161; *Crelle*, LXXXIII, 215, and is given also by Borchardt, *Crelle*, LXXXIII, 239.

In order to express the coefficients in terms of the k_s we require the connection* between these numbers and α, β, γ, δ. If the tetrahedron of reference is the fundamental tetrahedron (12, 34, 56), the six numbers

$$k_1 \quad k_2 \quad k_3 \quad k_4 \quad k_5 \quad k_6$$

are projectively related to the parameters of the nodes

$$(ab) \quad (ac) \quad (bc) \quad (ba) \quad (ca) \quad (cb)$$

lying on the section of the surface (p. 35)

$$\sqrt{[cd][db](dd)(cb)} + \sqrt{[bb][ad](ab)(bd)} + \sqrt{[da][cc](ca)(dc)} = 0$$

by the plane $(dd) = 0$,

and may therefore be taken equal to the values of

$$\frac{(ca)}{(bd)}, \quad = \frac{[bb][ad](ab)}{[da][cc](dc)}$$

at the nodes: these values are

$$\frac{[bb][ad][dd]}{[da][cc][aa]}, \quad \frac{[bb]}{[cc]}, \quad 0, \quad -\frac{[ad]}{[da]}, \quad -\frac{[dd]}{[aa]}, \quad \infty,$$

and their cross ratios are equal to the cross ratios of the corresponding k_s. Owing to the identical relations among the quadrics, this is only one among many ways in which the cross ratios of the k_s can be expressed. In particular we have

$$\frac{(k_3-k_5)(k_6-k_4)}{(k_3-k_4)(k_6-k_5)} = \frac{[dd][da]}{[aa][ad]} = \frac{2-A}{4},$$

whence
$$\frac{A}{2} = \frac{(k_3+k_4)(k_5+k_6) - 2(k_3k_4+k_5k_6)}{(k_3-k_4)(k_5-k_6)};$$

similarly
$$\frac{B}{2} = \frac{(k_5+k_6)(k_1+k_2) - 2(k_5k_6+k_1k_2)}{(k_5-k_6)(k_1-k_2)},$$

$$\frac{C}{2} = \frac{(k_1+k_2)(k_3+k_4) - 2(k_1k_2+k_3k_4)}{(k_1-k_2)(k_3-k_4)},$$

and we see that the vanishing of one of these coefficients is the condition for four nodes to form a harmonic range on the singular conic through them.

Lastly it may easily be verified that

$$D = 4\,\frac{k_1k_2(k_3+k_4-k_5-k_6)+k_3k_4(k_5+k_6-k_1-k_2)+k_5k_6(k_1+k_2-k_3-k_4)}{(k_1-k_2)(k_3-k_4)(k_5-k_6)},$$

when it is noticed that the vanishing of the numerator is the

* Cf. Bolza, *Math. Ann.* xxx, 478.

condition for the three pairs (k_1k_2), (k_3k_4), (k_5k_6) to form an involution. Thus all the coefficients in the quartic equation are expressed in terms of the coefficients of the quadratic complex of which the surface is the singular surface*.

The equation may be written as the sum of the squares of the fundamental quadrics multiplied by coefficients of which a typical one is

$$2k_1k_2k_3 + 2k_4k_5k_6 + \Sigma \pm k_\rho k_\sigma k_\tau;$$

the summation extends to twenty products and the positive sign is taken if the product of the symbols 123, $\rho\sigma\tau$, 135 according to the laws

$$0 = 11 = 22 = 33 = 44 = 55 = 66 = 123456,$$

is a three-figure symbol†.

§ 54. THE EQUATION REFERRED TO A ROSENHAIN TETRAHEDRON.

Take the linear forms belonging to any Rosenhain tetrad‡ for new coordinates, for example, those in the first column of the orthogonal matrix (p. 30),

$$x_1 = (dd), \quad x_2 = (ab), \quad x_3 = (bc), \quad x_4 = (ca).$$

The equation must be invariant for the operations represented by these symbols, but these transform the preceding coordinates into

dd	x_1,	x_2,	x_3,	x_4
ab	x_2,	$-x_1$,	$-x_4$,	x_3
bc	x_3,	x_4,	$-x_1$,	$-x_2$
ca	x_4,	$-x_3$,	x_2,	$-x_1$

respectively, showing that the terms $x_1^2x_4^2$ and $x_2^2x_3^2$ have the same coefficient, and that the term $x_1^2x_2x_3$ gives rise to the expression

$$x_1^2x_2x_3 + x_2^2x_1x_4 - x_3^2x_4x_1 - x_4^2x_3x_2 = (x_1x_2 - x_3x_4)(x_1x_3 + x_2x_4)$$

and so on. From the fact that all the corners of reference are nodes and all the faces tropes, we are able to write the equation in the form

$$u^2(x_1^2x_4^2 + x_2^2x_3^2) + v^2(x_2^2x_4^2 + x_3^2x_1^2) + w^2(x_3^2x_4^2 + x_1^2x_2^2)$$
$$+ 2vw(x_1x_2 - x_3x_4)(x_1x_3 + x_2x_4) + 2wu(\quad)(\quad) + 2uv(\quad)(\quad)$$
$$+ 2sx_1x_2x_3x_4 = 0,$$

or

$$x_4^2(u^2x_1^2 + v^2x_2^2 + w^2x_3^2 - 2vwx_2x_3 - 2wux_3x_1 - 2uvx_1x_2)$$
$$+ 2x_4\{vwx_1(x_2^2 - x_3^2) + wux_2(x_3^2 - x_1^2) + uvx_3(x_1^2 - x_2^2) + sx_1x_2x_3\}$$
$$+ (ux_2x_3 + vx_3x_1 + wx_1x_2)^2 = 0.$$

* Rohn, *Math. Ann.* XVIII, 142.
† Study, *Leipziger Berichte* (1892), XLVIII, 122.
‡ Cayley, *Coll. Papers*, VII, 126; *Crelle*, LXXIII, 292.

This equation may be deduced from the fact that Kummer's surface is the focal surface of the congruence of rays common to a tetrahedral complex and a linear complex. Using the notation of Chap. VI, we take the tetrahedral complex to be given by

$$p_{14}p_{23}/\alpha = p_{24}p_{31}/\beta = p_{34}p_{12}/\gamma,$$

equivalent to only one equation since

$$\alpha+\beta+\gamma=0,$$

and the linear complex to be

$$q_{14}p_{14}+q_{24}p_{24}+q_{34}p_{34}+q_{23}p_{23}+q_{31}p_{31}+q_{12}p_{12}=0,$$

where now the q_{rs} are any constants, not the coordinates of a line as in the case of Plücker's surface. Further we use the abbreviations $\xi_i = \Sigma q_{is}x_s$. Then the complex curve in the plane $\Sigma u_i x_i = 0$ is its intersection with the cone

$$\sqrt{\alpha u_1 x_1} + \sqrt{\beta u_2 x_2} + \sqrt{\gamma u_3 x_3} = 0.$$

If $u_i = \Sigma q_{is}x_s'$ then (x') is the null-point of the plane, and the two rays of the congruence through (x') are the two tangents from (x') to the complex curve. If (x') is on the focal surface, these two rays coincide and (x') must lie on the conic. Hence the equation of the focal surface is

$$\sqrt{\alpha\xi_1 x_1} + \sqrt{\beta\xi_2 x_2} + \sqrt{\gamma\xi_3 x_3} = 0.$$

Before expanding this, it is convenient to make a slight change in the coordinates, replacing x_1/x_4 by $x_1\sqrt{q_{24}q_{34}}/x_4\sqrt{q_{31}q_{12}}$ and so on, which has the effect of making the coefficients in the linear complex equal by pairs. We therefore write

$$q_{14}=q_{23}=l, \quad q_{24}=q_{31}=m, \quad q_{34}=q_{12}=n,$$

$$\xi_1 = nx_2 - mx_3 + lx_4,$$

$$\xi_2 = -nx_1 + lx_3 + mx_4,$$

$$\xi_3 = mx_1 - lx_2 + nx_4,$$

and then the equation

$$\sqrt{\alpha\xi_1 x_1} + \sqrt{\beta\xi_2 x_2} + \sqrt{\gamma\xi_3 x_3} = 0$$

gives, on expansion, the former equation, provided

$$u = l\alpha, \quad v = m\beta, \quad w = n\gamma,$$

$$s = u^2\alpha^{-1}(\gamma-\beta) + v^2\beta^{-1}(\alpha-\gamma) + w^2\gamma^{-1}(\beta-\alpha).$$

The same surface is obtained when $\alpha:\beta:\gamma$ are replaced by any other ratios satisfying the last preceding equation and

$$\alpha+\beta+\gamma=0.$$

Regarding these as trilinear equations of a plane cubic and a straight line, we see that there are three solutions $\alpha_1 : \beta_1 : \gamma_1$, $\alpha_2 : \beta_2 : \gamma_2$, and $\alpha_3 : \beta_3 : \gamma_3$ and it is easy to prove that

$$\frac{u^2}{\alpha_1\alpha_2} + \frac{v^2}{\beta_1\beta_2} + \frac{w^2}{\gamma_1\gamma_2} = 0$$

and two similar equations, and

$$\frac{u^2}{\alpha_1\alpha_2\alpha_3} = \frac{v^2}{\beta_1\beta_2\beta_3} = \frac{w^2}{\gamma_1\gamma_2\gamma_3}.$$

Each of these fractions may be equated to 1 since only the ratios $u : v : w$ are important, and then s can be put in the symmetrical form

$$\begin{aligned} 3s = &(\beta_1 - \gamma_1)(\beta_2 - \gamma_2)(\beta_3 - \gamma_3) \\ &+ (\gamma_1 - \alpha_1)(\gamma_2 - \alpha_2)(\gamma_3 - \alpha_3) \\ &+ (\alpha_1 - \beta_1)(\alpha_2 - \beta_2)(\alpha_3 - \beta_3). \end{aligned}$$

The corresponding sets of numbers (l_1, m_1, n_1), (l_2, m_2, n_2), (l_3, m_3, n_3) may be regarded as the direction cosines of three mutually orthogonal lines, and Kummer's surface can be written in three equivalent forms of which one is

$$\sqrt{l_2 l_3 x_1 (l_1 x_4 + n_1 x_2 - m_1 x_3)} + \sqrt{m_2 m_3 x_2 (m_1 x_4 + l_1 x_3 - n_1 x_1)} \\ + \sqrt{n_2 n_3 x_3 (n_1 x_4 + m_1 x_1 - l_1 x_2)} = 0.$$

A tetrahedron is a degenerate form of Kummer surface in which the nodes on each conic have coincided by pairs. Hence the set of six confocal congruences (p. 66), in which the quadratic complexes are cosingular and tetrahedral, reduces to *three*. It is immediately verified that the three linear complexes

$$l_i (p_{14} + p_{23}) + m_i (p_{24} + p_{31}) + n_i (p_{34} + p_{12}) = 0$$

are apolar, and are fundamental complexes for the surface. The others are

$$p_{14} = p_{23}, \quad p_{24} = p_{31}, \quad p_{34} = p_{12},$$

and one fundamental quadric is

$$x_1^2 + x_2^2 + x_3^2 + x_4^2 = 0,$$

so that the tangential equation has exactly the same form as the point equation.

The equation referred to a Göpel tetrad of tropes is (cf. p. 21 footnote)

$$[x^2 + y^2 + z^2 + t^2 + 2p(xt + yz) + 2q(yt + zx) + 2r(zt + xy)]^2 = 16sxyzt,$$

where $$s = p^2 + q^2 + r^2 - 2pqr - 1.$$

§ 55. NODAL QUARTIC SURFACES.

The equation of Kummer's surface has frequently appeared in the irrational form

$$\sqrt{xx'} + \sqrt{yy'} + \sqrt{zz'} = 0,$$

where $x=0, \dots z'=0$, are six planes. When these planes are arbitrary, the preceding equation evidently represents a quartic surface having six tropes and *fourteen* nodes, namely

$$(xyz),\ (x'yz),\ (xy'z),\ (xyz'),\ (xy'z'),\ (x'yz'),\ (x'y'z),\ (x'y'z'),$$

$$\left.\begin{matrix} x=0=x' \\ yy'=zz' \end{matrix}\right\},\quad \left.\begin{matrix} y=0=y' \\ zz'=xx' \end{matrix}\right\},\quad \left.\begin{matrix} z=0=z' \\ xx'=yy' \end{matrix}\right\}.$$

Conversely, the general fourteen-nodal surface can be expressed in this way. To prove this we notice that the sextic enveloping cone from each node must have thirteen double lines and hence must break up in one of the two following ways

$$3_1\ 1\ 1\ 1, \qquad 2\ 2\ 1\ 1,$$

where 3_1 denotes a cubic cone having one double line, and so on. Let m be the number of nodes of the first kind; then since each trope contains six nodes, the number of tropes is $(m+28)/6$, that is 5, 6, or 7. We take $x=0$, $y=0$, $z=0$ to be tropes meeting in a node of the first kind, and the quadric cone of tangents there to be

$$A \equiv x^2 + y^2 + z^2 - 2yz - 2zx - 2xy = 0.$$

Then if $\xi = 0$ is any plane not passing through this node, the surface has an equation of the form

$$A\xi^2 + 2B\xi + C = 0.$$

Since the enveloping cone breaks up in the assumed way,

$$B^2 - AC \equiv \mu xyz\theta,$$

where μ is a constant and $\theta = 0$ is a nodal cubic cone touching $A=0$ along three generators.

The nodal line on $\theta = 0$ may be taken arbitrarily, say

$$Y=0, \quad Z=0,$$

then $\theta=0$ has to satisfy six conditions and can contain only three arbitrary constants. It is sufficiently general to take

$$\theta \equiv (x-y-z)\,YZ - yZ^2 - zY^2,$$

for this satisfies all the conditions and contains implicitly three arbitrary constants, since Y and Z may be replaced by any linear functions of them. Introduce X where

$$X + Y + Z \equiv 0,$$

then

$$\theta \equiv xYZ + yZX + zXY.$$

The thirteen nodes other than $x=y=z=0$ lie on the lines

$$y=0=z, \quad z=0=x, \quad x=0=y,$$
$$X=Y=Z=0,$$
$$x=0=X, \quad y=0=Y, \quad z=0=Z,$$
$$x=0=yZ+zY, \quad y=0=zX+xZ, \quad z=0=xY+yX,$$

lying by sixes on the three cones

$$yZ+zY=0, \quad zX+xZ=0, \quad xY+yX=0.$$

Now there are at least two singular conics not passing through the first node and so we may take $\xi=0$ to be the trope containing the nodes on the first of these cones; then we must have

$$C \equiv \lambda\,(yZ+zY)^2,$$

and we may put $\lambda=1$ since the absolute value of ξ is undetermined. Then

$$B^2 \equiv \{(x-y-z)^2-4yz\}\,(yZ+zY)^2+\mu xyz\,(xYZ+yZX+zXY)$$

leading to $\mu=-4$ and

$$B \equiv yZ\,(x-y+z)-zY\,(x+y-z).$$

The equation of the surface is thus completely determined to be

$$(x^2+y^2+z^2-2yz-2zx-2xy)\,\xi^2$$
$$+2\,\{yZ\,(x-y+z)-zY\,(x+y-z)\}\,\xi+(yZ+zY)^2=0.$$

By introducing new linear expressions

$$\eta=\xi-Z, \quad \zeta=\xi+Y,$$

this equation becomes the rationalised form of

$$\sqrt{x\xi}+\sqrt{y\eta}+\sqrt{z\zeta}=0,$$

making evident the remaining tropes $\eta=0$, $\zeta=0$.

For this surface* to acquire a fifteenth node, the cubic cone must break up into a plane and a quadric cone, intersecting in lines passing through the fourteenth and fifteenth nodes. The cubic cone

$$x\,(\zeta-\xi)\,(\xi-\eta)+y\,(\xi-\eta)\,(\eta-\zeta)+z\,(\eta-\zeta)\,(\zeta-\xi)=0,$$

with a nodal line $\qquad \xi=\eta=\zeta,$

contains as part of itself the plane

$$l\xi+m\eta+n\zeta=0,$$

* For the equation of a thirteen-nodal surface which acquires additional nodes when one, two, or three constants are made to vanish see Kummer, *Berl. Abh.* (1866), p. 114, and Cayley, *Coll. Papers*, VII, 293.

where
$$l+m+n=0,$$
provided that, in consequence of these last two equations,
$$mnx+nly+lmz=0.$$
Hence an identical relation of the form
$$mnx+nly+lmz+k(l\xi+m\eta+n\zeta)=0$$
must exist. Now the quartic equation is unaltered if $x, y, z, \xi, \eta, \zeta$ are replaced by $px, qy, rz, qr\xi, rp\eta, pq\zeta$ respectively, p, q, r being arbitrary. Hence the preceding linear identity may have the more general form
$$mnpx+nlqy+lmrz+klqr\xi+kmrp\eta+knpq\zeta=0,$$
and, on writing it with undetermined coefficients
$$ax+by+cz+\alpha\xi+\beta\eta+\gamma\zeta=0,$$
the conditions become
$$a\alpha=b\beta=c\gamma.$$

There is of course one other linear identical relation among the six planes and the coefficients in this may be arbitrary. By what precedes, one of the new tropes is $ax+by+cz=0$, as may easily be verified directly, and similarly the three other new tropes are $\alpha\xi+by+cz=0$, $ax+\beta\eta+cz=0$, $ax+by+\gamma\zeta=0$, passing through the new node $ax=\alpha\xi$, $by=\beta\eta$, $cz=\gamma\zeta$.

For a sixteenth node to appear a second linear factor must separate out from the cubic θ, distinct from the former, but vanishing for the values $\xi=\eta=\zeta$. Exactly similar reasoning leads to a second identity of the form*
$$a'x+b'y+c'z+\alpha'\xi+\beta'\eta+\gamma'\zeta=0,$$
where
$$a'\alpha'=b'\beta'=c'\gamma',$$
and, as before, four new tropes appear, whose equations are $a'x+b'y+c'z=0$, $\alpha'\xi+b'y+c'z=0$, $a'x+\beta'\eta+c'z=0$, $a'x+b'y+\gamma'\zeta=0$, passing through the sixteenth node $a'x=\alpha'\xi$, $b'y=\beta'\eta$, $c'z=\gamma'\zeta$.

In the notation of p. 85, writing $x_1'=l_2l_3(n_1x_2-m_1x_3+l_1x_4)$ etc. the two relations are
$$l_2x_1+m_2x_2+n_2x_3+l_2^{-1}x_1'+m_2^{-1}x_2'+n_2^{-1}x_3'=0,$$
$$l_3x_1+m_3x_2+n_3x_3+l_3^{-1}x_1'+m_3^{-1}x_2'+n_3^{-1}x_3'=0.$$

* Jessop, *Quarterly Journal*, XXXI, 354.

CHAPTER IX.

SPECIAL FORMS OF KUMMER'S SURFACE.

§ 56. THE TETRAHEDROID.

We have seen in preceding chapters how the general 16_6 configuration depends upon six apolar complexes, and when these are given, is completely determined by a single element. Special configurations arise in two ways, either by specialising the linear complexes, as in the case of Plücker's surface, or by specialising the position of one of the elements. With the former case we are not here concerned* and confine our attention to the case when the set of apolar complexes is general, and consider the consequences of taking one node or trope of the surface in particular positions.

With respect to a single fundamental tetrahedron particular positions of a point are in a face, on an edge, or at a corner. If this tetrahedron is taken for reference, these three cases correspond to the vanishing of one, two, or three of the coordinates $(\alpha, \beta, \gamma, \delta)$ of one node.

We recall that the nodes and tropes are named after the operations which deduce them from $(\alpha, \beta, \gamma, \delta)$, and their equations are obtained by equating to zero the sixteen linear forms (p. 29)

$$\begin{bmatrix} (aa) & (ab) & (ac) & (ad) \\ (ba) & (bb) & (bc) & (bd) \\ (ca) & (cb) & (cc) & (cd) \\ (da) & (db) & (dc) & (dd) \end{bmatrix} \equiv \begin{bmatrix} \delta & \gamma & \beta & \alpha \\ \gamma & \delta & \alpha & \beta \\ \beta & \alpha & \delta & \gamma \\ \alpha & \beta & \gamma & \delta \end{bmatrix} \begin{bmatrix} x & -x & -x & x \\ -y & y & -y & y \\ -z & -z & z & z \\ t & t & t & t \end{bmatrix}.$$

The general case of a node lying in a face of a fundamental tetrahedron is given by $\delta = 0$, and then we see that the nodes

$$\begin{array}{lllll} (aa), & (ab), & (ac), & (ad) & \text{lie in } x=0, \\ (ba), & (bb), & (bc), & (bd) & \quad ,, \quad y=0, \\ (ca), & (cb), & (cc), & (cd) & \quad ,, \quad z=0, \\ (da), & (db), & (dc), & (dd) & \quad ,, \quad t=0, \end{array}$$

* See an exhaustive paper by Weiler, *Math. Ann.* VII, 145.

and the tropes pass by fours through the corners of reference. These sets of nodes are called Göpel tetrads, being corners of tetrahedra whose faces are not tropes, and the four groups make up a group-set, being either unchanged or interchanged by the group of sixteen operations. Thus, when $\delta = 0$, four tetrahedra belonging to a group-set become plane.

This special kind of Kummer surface is called a *Tetrahedroid**; it is characterised geometrically by the fact that the section by each of the faces of a certain tetrahedron is two conics, intersecting in four nodes.

The four tropes through any one corner intersect in six lines which, since they contain pairs of nodes, must lie in the faces. Hence each trope cuts three faces of the tetrahedron in three concurrent lines containing pairs of nodes; therefore the six nodes on any singular conic belong to an *involution* in which the chords joining corresponding points pass through a fundamental corner. This is characteristic of a tetrahedroid, that the six coefficients k_s in the complex of which it is the singular surface form three pairs of an involution. If the tetrahedron is (12, 34, 56), it is easy to see that the pairs of coefficients are $k_1 k_2$, $k_3 k_4$, $k_5 k_6$.

Since the sides of the quadrangle of nodes in any face pass through the corners, these form the common self-polar triangle of the two conics into which the section breaks up. From this fact it is easy to construct the equation of the surface, for it must be of the form

$$F(x^2, y^2, z^2, t^2) + \lambda xyzt = 0,$$

and after equating any coordinate to zero, F must break into factors. Under these conditions the equation represents a surface which touches each of the planes of reference at four points; if one of these is a node λ must be zero and then the other fifteen points of contact are nodes also. The conditions for F show that the equation may be written in the form

$$\begin{vmatrix} 0 & h & g & l & x^2 \\ h & 0 & f & m & y^2 \\ g & f & 0 & n & z^2 \\ l & m & n & 0 & t^2 \\ x^2 & y^2 & z^2 & t^2 & 0 \end{vmatrix} = 0.$$

By replacing $(fmn)^{\frac{1}{4}}x$, $(gnl)^{\frac{1}{4}}y$, $(hlm)^{\frac{1}{4}}z$, $(fgh)^{\frac{1}{4}}t$ by new coordi-

* Cayley, *Coll. Papers*, I, 302; *Liouville* (1846), XI, 291.

nates x, y, z, t and putting $\alpha = (fl)^{\frac{1}{2}}$, $\beta = (gm)^{\frac{1}{2}}$, $\gamma = (hn)^{\frac{1}{2}}$, this takes the form

$$\begin{vmatrix} 0 & \gamma^2 & \beta^2 & \alpha^2 & x^2 \\ \gamma^2 & 0 & \alpha^2 & \beta^2 & y^2 \\ \beta^2 & \alpha^2 & 0 & \gamma^2 & z^2 \\ \alpha^2 & \beta^2 & \gamma^2 & 0 & t^2 \\ x^2 & y^2 & z^2 & t^2 & 0 \end{vmatrix} = 0,$$

or on expansion

$$\alpha^2\beta^2\gamma^2 (x^4 + y^4 + z^4 + t^4) + \alpha^2 (\alpha^4 - \beta^4 - \gamma^4)(x^2t^2 + y^2z^2)$$
$$+ \beta^2 (\beta^4 - \gamma^4 - \alpha^4)(y^2t^2 + z^2x^2) + \gamma^2 (\gamma^4 - \alpha^4 - \beta^4)(z^2t^2 + x^2y^2) = 0,$$

which is what the general equation (p. 81) becomes when $\delta = 0$.

In order to exhibit the conics in the planes of reference put $(\alpha^4 - \beta^4 - \gamma^4)/\beta^2\gamma^2 = (\mu + \nu)/\sqrt{\mu\nu}$, etc. and replace x^2 by $x^2\sqrt{\mu\nu}$, y^2 by $y^2\sqrt{\nu\lambda}$, z^2 by $z^2\sqrt{\lambda\mu}$; then the equation becomes

$$(x^2 + y^2 + z^2)(\mu\nu x^2 + \nu\lambda y^2 + \lambda\mu z^2)$$
$$+ t^2 \{(\mu + \nu) x^2 + (\nu + \lambda) y^2 + (\lambda + \mu) z^2\} + t^4 = 0,$$

and finally on putting $\lambda = -a^{-2}$, $\mu = -b^{-2}$, $\nu = -c^{-2}$, $t = 1$, we get the ordinary equation of the wave surface

$$(x^2 + y^2 + z^2)(a^2x^2 + b^2y^2 + c^2z^2)$$
$$- \{a^2(b^2 + c^2) x^2 + b^2(c^2 + a^2) y^2 + c^2(a^2 + b^2) z^2\} + a^2b^2c^2 = 0.$$

The two points of contact of a common tangent of the conics in a face of the tetrahedron lie on the same singular conic and are the double points of the involution of nodes on that conic.

If a given linear complex is apolar to the complex of polar lines of its rays with respect to a variable quadric having a fixed self-polar tetrahedron, the envelope of the quadric is a tetrahedroid. If the given complex is special, the variable quadric touches eight fixed lines and the tetrahedroid degenerates into a repeated quadric.

The intersections of corresponding surfaces in an involution of quadrics inscribed in a common developable generate a tetrahedroid.

When the determining node is taken on an edge of a fundamental tetrahedron, the surface becomes a scroll of which a typical equation is *

$$\alpha\beta(k^2\alpha^2 - \beta^2)(x^2t^2 + y^2z^2) + \alpha\beta(\alpha^2 - k^2\beta^2)(y^2t^2 + z^2x^2) + 2k(\alpha^4 - \beta^4)\,xyzt = 0.$$

§ 57. MULTIPLE TETRAHEDROIDS.

Double tetrahedroid.

The trope, from which all the remaining singularities are obtained, passes through one corner of each of *two* fundamental tetrahedra. These must not belong to the same desmic system, else a third corner would be collinear with the other two and so lie in the trope. Hence the two tetrahedra have two edges

* Rohn, *Math. Ann.* XVIII, 156.

common, and without loss of generality we may take the corners to be

$$(0,\ 0,\ 0,\ 1) \text{ and } (0,\ 1, -1,\ 0).$$

This gives $\delta = 0$ and $\beta = \gamma$, and the equation of the tetrahedroid becomes*

$$(x^2 - t^2)^2 + (y^2 - z^2)^2 + \frac{\alpha^4}{\beta^4}(x^2t^2 + y^2z^2) - \frac{\alpha^2}{\beta^2}(x^2 + t^2)(y^2 + z^2).$$

As in the case of the single tetrahedroid, when a trope contains a fundamental corner it cuts the three concurrent faces in lines joining pairs of nodes. The corner is therefore a centre of perspective for two triangles formed by the six nodes in the trope. In the present case the two corners lie one on each of the common edges of the two tetrahedra, and the line joining them lies in a face of each and therefore joins two nodes. Hence in the case of a double tetrahedroid every trope passes through two fundamental corners which are collinear with two nodes, and a second trope passes through the same two corners. The arrangement of nodes on any conic is projectively equivalent to that shown in fig. 8, and the *fundamental sextic* $\Pi(k - k_s)$ may be linearly transformed into the form $Ak^5 + Bk^3 + Ck$. Corresponding to the arbitrary numbering of the figure the two tetrahedra are (14, 25, 36) and (12, 36, 45) having the edges (36) common, and the triangles 153, 246 are in two-fold perspective, and so also are the triangles 156, 243. By an imaginary projection we may take 3 and 6 to be the circular points at infinity and then 24 and 15 become diameters.

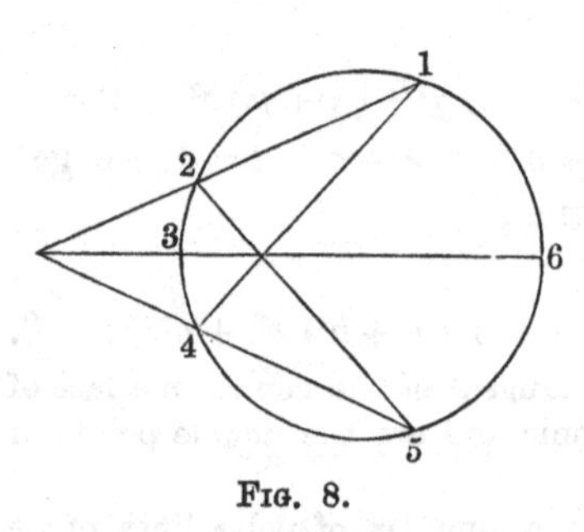

FIG. 8.

Triple tetrahedroid.

Each trope passes through three fundamental corners. If these are collinear, it is sufficient to make the trope pass through a corner of each of two desmic tetrahedra, for example (0, 0, 0, 1) and (1, 1, −1, 1) so that

$$\delta = 0 \text{ and } \alpha + \beta - \gamma = 0.$$

Then the equation of the surface is

$$x^4 + y^4 + z^4 + t^4 - \frac{4\alpha^2 + 2\alpha\beta + 2\beta^2}{\beta(\alpha+\beta)}(x^2t^2 + y^2z^2)$$
$$- \frac{2\alpha^2 + 2\alpha\beta + 4\beta^2}{\alpha(\alpha+\beta)}(y^2t^2 + z^2x^2) + \frac{4\alpha^2 + 6\alpha\beta + 4\beta^2}{\alpha\beta}(z^2t^2 + x^2y^2) = 0.$$

* Rohn, *Leipziger Berichte* (1884), XXXVI, 10.

The collinearity of the three centres of perspective shows that they are the three points on a Pascal line. Projecting them to infinity and the conic into a circle we have fig. 9, and the fundamental sextic may be taken to be of the form $Ak^6 + Bk^3 + C$. The tetrahedra corresponding to the figure are (12, 36, 45), (34, 25, 16), (56, 14, 23), of the same desmic system and the triangles 135, 246 are in threefold perspective.

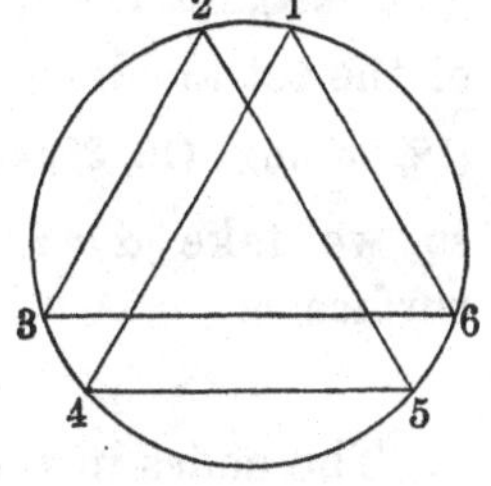

FIG. 9.

Quadruple tetrahedroid.

Any three corners which are not collinear are necessarily coplanar with a fourth, for of any three tetrahedra two belong to a desmic system or else all three have two common edges. Start with one tetrahedron (14, 25, 36) and take two others each having two edges in common with it, for example (12, 36, 45) and (23, 14, 56); these two belong to a desmic system of which the third member is (34, 16, 25). It is sufficiently general to take the three corners to be

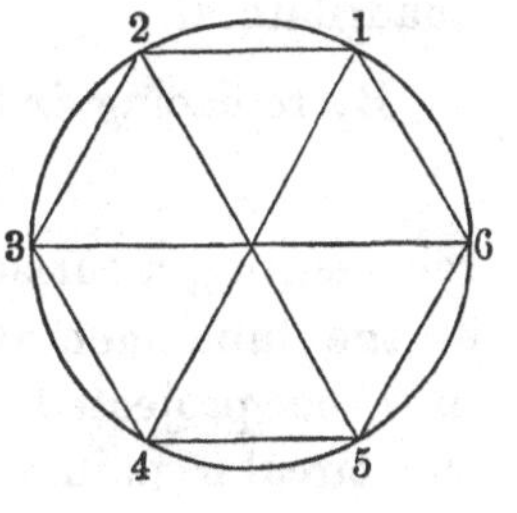

FIG. 10.

(0, 0, 0, 1), (0, 1, −1, 0), (1, 0, −1, 0),

and then $\delta = 0, \quad \alpha = \beta = \gamma,$

and the equation is

$$x^4 + y^4 + z^4 + t^4 - x^2t^2 - y^2t^2 - z^2t^2 - y^2z^2 - z^2x^2 - x^2y^2 = 0.$$

The figure of coplanar nodes is projected into a regular hexagon and the triangles 135, 246 formed by alternate corners are in fourfold perspective. The fundamental sextic when linearly transformed into its simplest form is $k^6 + 1$.

*Sextuple tetrahedroid**.

It can be shown that the case of only five coplanar corners does not arise, but that there exist planes passing through six corners, and that the tetrahedra to which they belong are a set having pairs of edges in common with the same tetrahedron. For example the plane

$$ix + y + z + t = 0$$

* Segre, *Leipziger Berichte* (1884), XXXVI, 132.

contains the corners

$(i, 0, 0, 1)$, $(0, 1, -1, 0)$, $(0, 1, 0, -1)$, $(i, 0, 1, 0)$, $(0, 0, 1, -1)$, $(i, 1, 0, 0)$

of the tetrahedra

(12, 35, 46), (12, 36, 45), (34, 16, 25), (34, 15, 26), (56, 14, 23), (56, 13, 24),

so we take $\alpha = i$, $\beta = \gamma = \delta = 1$. Then the equation of the surface is

$$x^4 + y^4 + z^4 + t^4 + 4ixyzt = 0.$$

The nodes in the trope $ix + y + z + t = 0$ are

$(1, -1, 1, -i), (1, 1, -1, -i), (1, 1, -i, -1), (-1, 1, i, -1), (-1, i, 1, -1), (1, -i, 1, -1)$,

and if they are denoted by 1, 2, 3, 4, 5, 6 respectively, the names of the tetrahedra show the six ways in which the lines joining them are concurrent. The six corners lie by threes on four lines and the corresponding tetrahedra belong to four desmic systems. The pairs of nodes 12, 34, 56 lie on the three diagonals of this quadrilateral.

By replacing ix by x we obtain a real equation

$$x^4 + y^4 + z^4 + t^4 + 4xyzt = 0,$$

representing a surface with four real nodes. The canonical form of the fundamental sextic is $k(k^4 - 1)$, and six coplanar nodes may be projected into the corners of a square together with the circular points at infinity.

The whole configuration of nodes is projected from a corner of reference into sixteen points lying by fours on twelve lines *.

§ 58. BATTAGLINI'S HARMONIC COMPLEX.

The tetrahedroid is a special form of Kummer's surface due to the fact that the six coefficients in the quadratic complex

$$k_1x_1^2 + k_2x_2^2 + k_3x_3^2 + k_4x_4^2 + k_5x_5^2 + k_6x_6^2 = 0$$

belong to an involution†. If the fundamental tetrahedron to which the surface is specially related is (12, 34, 56) the condition for involution is

$$\begin{vmatrix} k_1k_2 & k_1 + k_2 & 1 \\ k_3k_4 & k_3 + k_4 & 1 \\ k_5k_6 & k_5 + k_6 & 1 \end{vmatrix} = 0.$$

* Kantor, *American Journal*, XIX, 86.

† Sturm, *Liniengeometrie*, III, 328; Battaglini, *Giornale di Matematiche*, VI and VII; Schur, *Mathematische Annalen*, XXI, 515.

By making a suitable linear transformation of the k_s, thereby producing another complex of the same cosingular family, we may arrange that

$$k_1 + k_2 = 0, \quad k_3 + k_4 = 0, \quad k_5 + k_6 = 0,$$

so that the complex has the form

$$k_1 (x_1^2 - x_2^2) + k_3 (x_3^2 - x_4^2) + k_5 (x_5^2 - x_6^2) = 0,$$

or, in Plücker's coordinates,

$$k_1 (p_{14}^2 + p_{23}^2) + k_3 (p_{24}^2 + p_{31}^2) + k_5 (p_{34}^2 + p_{12}^2) = 0,$$

and is distinguished, in this notation, by the absence of product terms.

This complex consists of all the lines cutting two quadrics harmonically, for, using point coordinates x_1, x_2, x_3, x_4, and taking the quadrics to be

$$a_1 x_1^2 + a_2 x_2^2 + a_3 x_3^2 + a_4 x_4^2 = 0,$$
$$b_1 x_1^2 + b_2 x_2^2 + b_3 x_3^2 + b_4 x_4^2 = 0,$$

the condition for a line (p) to cut them in two pairs of harmonically conjugate points is

$$\Sigma (a_r b_s + a_s b_r) p_{rs}^2 = 0,$$

and by a slight change of coordinates this can be reduced to the preceding form.

Since the line equation of the quadric

$$\Sigma a_s x_s^2 + \lambda \Sigma b_s x_s^2 = 0,$$

is

$$\Sigma (a_r + \lambda b_r)(a_s + \lambda b_s) p_{rs}^2 = 0,$$

it follows that two quadrics of the pencil obtained by varying λ touch any given line. If their parameters are λ and λ' and the line is a ray, the equation of the complex expresses the condition

$$\lambda + \lambda' = 0.$$

Singular rays belong to the tetrahedral complex

$$\Sigma (a_1 a_4 b_2 b_3 + a_2 a_3 b_1 b_4) p_{14} p_{23} = 0,$$

and it is easily proved that when this condition is satisfied, all the polar lines of (p) with respect to the pencil of quadrics intersect and generate a cone of rays of this tetrahedral complex. The two polar lines of a singular ray (p) with respect to the two quadrics which touch it pass through the points of contact and lie in the tangent planes, which must coincide since the two polar lines intersect. Hence the congruence of singular rays consists of all the generators of the developables circumscribing the pairs of quadrics

$$\Sigma a_s x_s^2 \pm \lambda \Sigma b_s x_s^2 = 0$$

for different values of λ.

When the complex is given, there is an infinite number of pairs of quadrics that may be used to define it. Let the complex be written

$$\Sigma A_{rs} p_{rs}^2 = 0,$$

where
$$A_{rs} = a_r b_s + a_s b_r.$$

Let
$$\Delta = a_1 a_2 a_3 a_4, \quad \Theta = \Delta \Sigma b_s / a_s,$$

then
$$\Theta = a_3 a_4 A_{12} + a_1 a_2 A_{34},$$

and
$$A_{34}(a_1 a_2)^2 - \Theta(a_1 a_2) + A_{12}\Delta = 0.$$

Write $4\Delta = \sigma\Theta^2$, then

$$2A_{34} a_1 a_2 = \Theta(1 + \sqrt{1 - \sigma A_{12} A_{34}}).$$

Similarly
$$2A_{12} a_3 a_4 = \Theta(1 - \sqrt{1 - \sigma A_{12} A_{34}})$$

and so on, giving the ratios of the a_s in terms of an arbitrary parameter σ. When they are found, the b_s are uniquely determined.

The quartic intersection of the quadrics A, $\Sigma a_s x_s^2 = 0$, and B, $\Sigma b_s x_s^2 = 0$, lies on the singular surface, for the complex cone at any point of it breaks up into the two tangent planes. Again the intersection of A and the reciprocal of A with respect to B lies on the singular surface, for the polar plane of any point of it with respect to B cuts A in two straight lines and the complex cone consists of the planes through them. The quadric surface C, $\Sigma c_s x_s^2 = 0$, where

$$c_s = a_s + \mu b_s^2 / a_s,$$

passes through this second quartic for all values of μ, and one value of μ exists for which C is one of a pair of quadrics which can define the complex. If $\Theta' = b_1 b_2 b_3 b_4 \Sigma a_s / b_s$ this value of μ is Θ/Θ' and we find

$$c_1 = \frac{A_{12} A_{13} A_{14}}{a_1 \Theta'}, \text{ etc.}$$

whence
$$2A_{34} c_1 c_2 = \frac{\Pi A_{rs}}{\Delta \Theta'^2} 2A_{12} a_3 a_4$$
$$= \Theta''(1 - \sqrt{1 - \sigma A_{12} A_{34}})$$

say, and
$$2A_{12} c_3 c_4 = \Theta''(1 + \sqrt{1 - \sigma A_{12} A_{34}}),$$

so that C corresponds to the same value of the parameter σ but to a different arrangement of the signs of the radicals.

In a similar way the quadric D, $\Sigma d_s x_s^2 = 0$, where

$$d_1 = \Theta b_1 + \Theta' a_1^2 / b_1 = A_{12} A_{13} A_{14} / b_1, \text{ etc.}$$

cuts B in a quartic curve lying on the singular surface.

The harmonic complex can be defined in a reciprocal manner as the assemblage of lines from which the tangent planes to two quadrics form two harmonically conjugate pairs. Taking the two tangential equations to be

$$\alpha_1 u_1^2 + \alpha_2 u_2^2 + \alpha_3 u_3^2 + \alpha_4 u_4^2 = 0,$$

$$\beta_1 u_1^2 + \beta_2 u_2^2 + \beta_3 u_3^2 + \beta_4 u_4^2 = 0,$$

the complex is

$$\Sigma\, (\alpha_3\beta_4 + \alpha_4\beta_3)\, p_{12}^2 = 0.$$

On putting $\alpha_s = c_s^{-1}$, $\beta_s = d_s^{-1}$ this becomes the same as before: accordingly the quadrics C and D can be used to define the complex tangentially.

The harmonic complex includes among its rays all the generators of the quadrics in terms of which it is defined: and conversely if a quadratic complex contains both sets of generators of a quadric it must be harmonic.

The complexes $\Sigma A_{12} p_{12}^2 = 0$, $\Sigma A_{34}^{-1} p_{12}^2 = 0$, have the same singular surface, and are the only harmonic complexes in the cosingular family.

The series of surfaces $\Sigma a_{s,n} x_s^2 = 0$, $\Sigma b_{s,n} x_s^2 = 0$, where, for every n,

$$a_{r,n} b_{s,n} + a_{s,n} b_{r,n} = A_{rs},$$

$$a_{s,n+1} = b_{s,n} a_{1,n} a_{2,n} a_{3,n} a_{4,n} \Sigma b_{s,n}/a_{s,n} + (a^2_{s,n}/b_{s,n})\, b_{1,n} b_{2,n} b_{3,n} b_{4,n} \Sigma a_{s,n}/b_{s,n}$$

$$= b_{s,n}\Theta_n + \frac{a_{s,n}^2}{b_{s,n}}\Theta_n', \text{ say,}$$

terminates after six members if

$$A_{14}^2 A_{23}^2 + A_{24}^2 A_{31}^2 + A_{34}^2 A_{12}^2$$
$$- 2A_{24}A_{31}A_{34}A_{12} - 2A_{34}A_{12}A_{14}A_{23} - 2A_{14}A_{23}A_{24}A_{31} = 0.$$

Express this result in terms of the invariants of the first two surfaces.

A common tangent plane of the two quadrics $S_1 \pm \lambda S_2 = 0$ is a singular plane of the complex of lines cutting $S_1 = 0$, $S_2 = 0$ harmonically. The singular ray joins the points of contact, and cuts the singular surface on $S_1 + \mu S_2 = 0$, if $S_1 - \mu S_2 = 0$ is the third surface of the pencil which touches the plane.

The singular surface of the complex $\Sigma A_{12} p_{12}^2 = 0$ is generated by the intersection of the surfaces

$$\lambda x_1^2 + A_{12} x_2^2 + A_{13} x_3^2 + A_{14} x_4^2 = 0,$$

$$A_{12} x_1^2 + \mu x_2^2 + A_{23} x_3^2 + A_{24} x_4^2 = 0,$$

where λ and μ are connected by a certain homographic relation.

§ 59. LIMITING FORMS.

When one node lies on a fundamental quadric the quartic degenerates into that quadric, repeated; for, since the quadric is invariant for the group of linear transformations which derive the nodes from any one of them, in this case all the nodes and therefore all the singular conics lie on a surface of the second order. This also follows easily from the equation of the surface,

$$\sqrt{\alpha x\xi} + \sqrt{\beta y\eta} + \sqrt{\gamma z\zeta} = 0,$$

(cf. p. 35), for when $\alpha = 0$ the rationalised form reduces to

$$(\beta y\eta - \gamma z\zeta)^2 = 0.$$

Since the polar plane of a node is a trope containing six other nodes, here each trope is a tangent plane to the quadric and contains seven nodes, the additional one being at the point of contact. Hence the nodes are at the intersections of four generators of one system with four of the other system.

When one node is very near the fundamental quadric, the whole quartic surface lies near the quadric, and consists of thin pieces joined together at the nodes. We may suppose the quartic to be derived from the quadric by each point of the latter separating into two, which may be real and distinct, or conjugate imaginaries. It is evident that the tropes touch the thin pieces of surface near their edges, just as, in two dimensions, a bitangent touches a thin branch of a curve near the points where it is folded on itself. If we regard the points of the quartic as determined by one node and varying continuously with it, we may distinguish, on the quadric, regions of points which are just going to separate into real and into imaginary pairs of points of the quartic; these regions are bounded by the singular conics, and so we have the figure

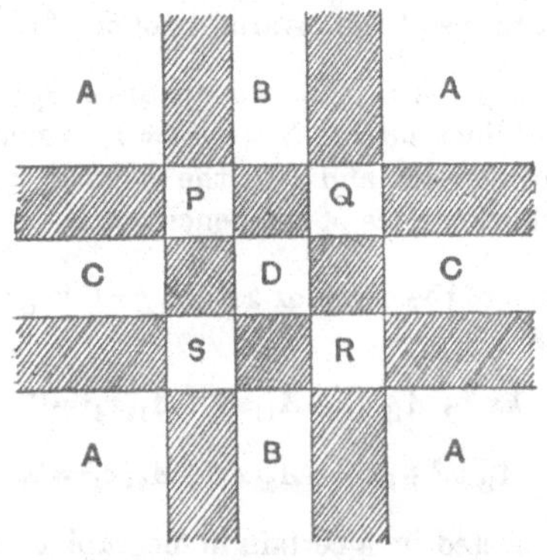

FIG. 11.

in which either the shaded or the unshaded portions may be considered as the real surface. By regarding the generators as closed curves we see that the surface consists of eight four-cornered pieces, and that each of four A, B, C, D is attached to each of the other four P, Q, R, S at one node.

The general Kummer surface can be obtained from a degenerate surface of this kind by a continuous variation of its points, without passing through any other degenerate form; for the nodes can be varied so as to avoid the edges of the fundamental tetrahedra. For this reason the limiting form has an important bearing on the topology of the general surface. In particular, the incidences of nodes and tropes remain the same during the variation, except that in the limiting form there is an extra node in each trope, namely the point of contact. *Hence the preceding figure, which is an actual representation of the limiting surface, is also an incidence diagram for the general surface**. Conversely, the rows and columns of the diagram of incidences can be regarded as generators of a quadric which is a limiting form of Kummer's surface.

Let the fundamental quadric be (135, 246); the directrices (13)(35)(51) are generators of one system and (24)(46)(62) are generators of the other system. One node, 0, may be taken arbitrarily on the surface; then the two generators through 0 together with their harmonic conjugates with respect to the pairs of directrices are the eight generators whose intersections give the nodes. If the harmonic conjugates are named after the directrices, and if the intersection of 13 and 24 is called 56, and so on, we have, as on p. 17, the incidence diagram

0	46	62	24
35	12	14	16
51	32	34	36
13	52	54	56.

* Rohn, *Math. Ann.* xv, 339; Klein, *Evanston Lectures* (1893), p. 29.

CHAPTER X.

THE WAVE SURFACE.

§ 60. DEFINITION OF THE SURFACE.

The Wave surface is a special form of the Tetrahedroid, and the latter is derived by a general linear transformation from the former, so that the two surfaces have the same projective features. In addition, the Wave surface possesses metrical properties of great interest, and is probably the best known example of Kummer's surface on account of its connection with the physical world*.

The specialisation is that the tetrahedron whose faces contain the nodes by fours is a rectangular frame of reference, one face being the plane at infinity and one of the conics in that face being the imaginary circle at infinity. From this it follows that one of the two conics in each of the other faces is a circle.

The details of the transformation from the general Tetrahedroid are given on p. 91 and the equation of the Wave surface is obtained in the form

$$(x^2+y^2+z^2)(a^2x^2+b^2y^2+c^2z^2) \\ -a^2(b^2+c^2)x^2-b^2(c^2+a^2)y^2-c^2(a^2+b^2)z^2+a^2b^2c^2=0,$$

which is equivalent to

$$\frac{a^2x^2}{a^2-r^2}+\frac{b^2y^2}{b^2-r^2}+\frac{c^2z^2}{c^2-r^2}=0,$$

where

$$r^2=x^2+y^2+z^2.$$

It is convenient to take this equation as a starting point and to deduce properties of the surface from it. If r is regarded as a constant, the equation represents the cone which is reciprocal to the cone

$$(a^{-2}-r^{-2})x^2+(b^{-2}-r^{-2})y^2+(c^{-2}-r^{-2})z^2=0$$

* Fresnel, *Œuvres Complètes*, II, 261; Preston, *Theory of Light*, p. 260.

passing through the intersection of the sphere

$$x^2 + y^2 + z^2 = r^2$$

and the ellipsoid $\quad a^{-2}x^2 + b^{-2}y^2 + c^{-2}z^2 = 1.$

Hence points on the Wave surface are obtained by taking any central section of the ellipsoid and measuring lengths equal to the semi-axes along the perpendicular at the centre, in both directions. If we suppose that $a > b > c$, all the points of the surface are at finite distances from the origin varying from c to a.

§ 61. APSIDAL SURFACES.

Let Q be any point of a surface S and let OM be the perpendicular from any point O on the tangent plane at Q. Draw QT and OR at right angles to the plane QOM; then Q is an *apse*

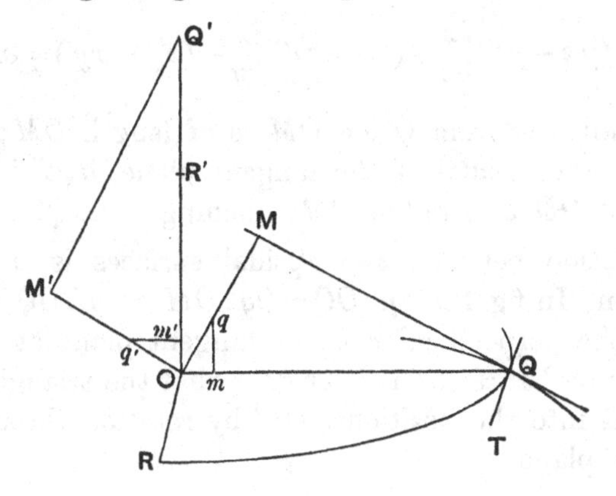

Fig. 12.

of the section by the plane ROQ with respect to the point O, for the tangent line QT is perpendicular to OQ. If OQ' is erected perpendicular to ROQ and of length equal to the apsidal radius OQ, the locus of Q' is called the *apsidal surface* S'.

A sphere with centre O and radius OQ cuts S in a curve to which QT is the tangent at Q; this curve is projected from O by a cone having ROQ for a tangent plane. Hence as OQ describes this cone, OQ' describes the reciprocal cone which cuts the sphere in a curve lying on the apsidal surface S'. We thus have a method of generating S' by means of spherical curves corresponding to spherical sections of S, and the relation between the surfaces is mutual. From this it follows that Q' is an apse of the section of S' by $Q'OR$, and hence the tangent plane at Q' is perpendicular to $Q'OQ$.

The apsidal surface S' may also be regarded as the focal surface of a congruence of circles having a common centre O; for the circle whose axis and radius is OQ touches S' at Q'. Corresponding to a point of S near Q in the plane QOQ' there is a circle touching S' near Q'. From this it is easily seen that the tangent plane at Q' is obtained by rotating the tangent plane at Q through a right angle about OR.

The fact that the relation between the tangent planes is independent of the curvatures shows that each of two apsidal surfaces is derived from the other by a *contact transformation**. This may easily be verified from the equations which define Q'

$$x'^2 + y'^2 + z'^2 = x^2 + y^2 + z^2,$$

$$x'x + y'y + z'z = 0,$$

$$(y'z - yz')\frac{\partial z}{\partial x} + (z'x - zx')\frac{\partial z}{\partial y} - (x'y - xy') = 0.$$

The perpendicular from Q' on OM is of length OM; whence it follows that the apsidal of the tangent plane MQT is a cylinder whose axis is OM and radius OM, touching S' at Q'.

The relation between two apsidal surfaces is unaltered by reciprocation. In fig. 12, $Om \, . \, OQ = Oq \, . \, OM = Om' \, . \, OQ' = Oq' \, . \, OM'$, and Om is the perpendicular on the tangent plane at the point q on the reciprocal surface; it is evident that the triangle qOm may be displaced into the position $q'Om'$ by rotation through a right angle in its plane.

§ 62. SINGULARITIES OF THE WAVE SURFACE.

Since the Wave surface is the apsidal of an ellipsoid with respect to its centre, many of its properties may be deduced in an elementary manner from those of the ellipsoid. In fig. 12, Q and R are apses of a central section and the perpendicular at O meets the wave surface at points Q' and R' such that $OQ' = OQ$ and $OR' = OR$, and at the images of these points in O. As the plane QOR varies, the points Q' and R' describe different *sheets* which meet only when $OQ = OR$, that is, when the central section of the ellipsoid is a circle. In this case every point on the section is an apse and there are an infinite number of tangent planes at Q'

* Lie-Scheffers, *Berührungstransformationen.*

which is therefore a node. Hence the two sheets are connected at four nodes lying on the perpendiculars to the circular sections of the ellipsoid: their coordinates are

$$\pm c(a^2-b^2)^{\frac{1}{2}}(a^2-c^2)^{-\frac{1}{2}},\quad 0,\quad \pm a(b^2-c^2)^{\frac{1}{2}}(a^2-c^2)^{-\frac{1}{2}}.$$

Reciprocally, the curve of contact with the ellipsoid of a circumscribing circular cylinder gives on the apsidal surface a curve at every point of which the tangent plane is the same.

The reciprocal singularities may be obtained by considering the reciprocal ellipsoid; and we infer that there are four real tropes

$$\pm(a^2-b^2)^{\frac{1}{2}}x \pm (b^2-c^2)^{\frac{1}{2}}z = b(a^2-c^2)^{\frac{1}{2}}.$$

It is easy to verify that the curves of contact are circles.

The section by the plane $x=0$ consists of the two conics

$$(y^2+z^2-a^2)(b^2y^2+c^2z^2-b^2c^2)=0,$$

namely a circle of radius a surrounding an ellipse of semi-axes b and c. The section by $y=0$ is

$$(z^2+x^2-b^2)(c^2z^2+a^2x^2-c^2a^2)=0,$$

a circle of radius b and an ellipse, intersecting in the four nodes. The section by $z=0$ is

$$(x^2+y^2-c^2)(a^2x^2+b^2y^2-a^2b^2)=0,$$

a circle of radius c surrounded by an ellipse; and lastly, the plane at infinity cuts the surface in the two imaginary conics

$$(x^2+y^2+z^2)(a^2x^2+b^2y^2+c^2z^2)=0.$$

The symmetry with respect to the planes of reference is evident from the fact that only squares of the coordinates occur in the equation. By drawing quadrants of the preceding conics, an idea of the shape of the surface may be obtained. Fig. 13 shows one node and the trace of one trope.

The shape may be more completely realised by tracing the series of sphero-conics cut out by concentric spheres. The projection of the intersection with a sphere of radius r upon the plane $z=0$ is

$$\frac{a^2-c^2}{a^2-r^2}x^2+\frac{b^2-c^2}{b^2-r^2}y^2=c^2,$$

which is a hyperbola for the outer sheet ($a>r>b$), and an ellipse for the inner sheet $(b > r > c)$. Again, projecting on the plane $y = 0$ we get the family of ellipses

$$\frac{a^2 - b^2}{a^2 - r^2} x^2 + \frac{b^2 - c^2}{r^2 - c^2} z^2 = b^2,$$

having for envelope the four lines

$$\pm (a^2 - b^2)^{\frac{1}{2}} x \pm (b^2 - c^2)^{\frac{1}{2}} z = b (a^2 - c^2)^{\frac{1}{2}}.$$

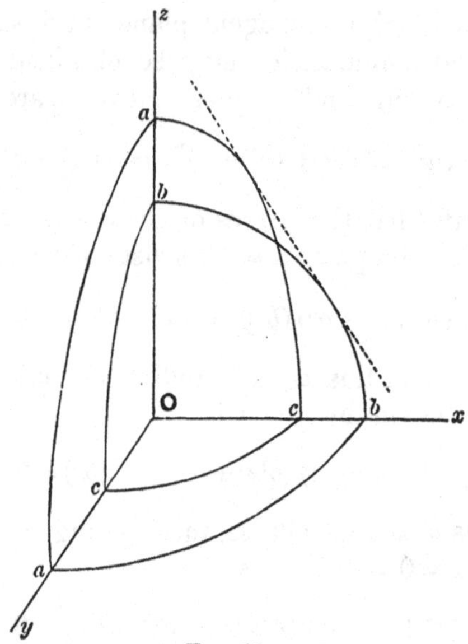

FIG. 13.

§ 63. PARAMETRIC REPRESENTATION.

It is convenient* to replace a^2 by a, b^2 by b, and c^2 by c. Let $P, (\xi, \eta, \zeta)$, be the end of the diameter of the ellipsoid conjugate to QOR (p. 101), then if $OQ^2 = \lambda$ and $OR^2 = \mu$, λ and μ are the parameters of the confocals through P and we have

$$(a - b)(a - c)\,\xi^2 = a (a - \lambda)(a - \mu),$$
$$(b - c)(b - a)\,\eta^2 = b (b - \lambda)(b - \mu),$$
$$(c - a)(c - b)\,\zeta^2 = c (c - \lambda)(c - \mu).$$

If p, p_1, p_2 are the central perpendiculars on the tangent planes at P to the ellipsoid and the two confocals,

$$\lambda \mu p^2 = abc,$$
$$\lambda (\lambda - \mu)\, p_1^2 = (a - \lambda)(b - \lambda)(c - \lambda),$$
$$\mu (\mu - \lambda)\, p_2^2 = (a - \mu)(b - \mu)(c - \mu).$$

* Darboux, *Comptes Rendus*, XCVII, 1039; Cayley, XIII, 238.

The point Q' of the wave surface is obtained by measuring a length $\lambda^{\frac{1}{2}}$ in the direction

$$p\xi/a, \quad p\eta/b, \quad p\zeta/c,$$

and is therefore given by

$$(a-b)(a-c)\,x^2 = bc\mu^{-1}(a-\lambda)(a-\mu),$$
$$(b-c)(b-a)\,y^2 = ca\mu^{-1}(b-\lambda)(b-\mu),$$
$$(c-a)(c-b)\,z^2 = ab\mu^{-1}(c-\lambda)(c-\mu),$$

expressing the coordinates of any point in terms of two parameters λ and μ. Conversely the parameters are expressed in terms of the coordinates by

$$x^2 + y^2 + z^2 = \lambda,$$
$$ax^2 + by^2 + cz^2 = abc\mu^{-1}.$$

In order to express x, y and z as *uniform* functions we must replace λ and μ by elliptic functions of new parameters p and q*. For the outer sheet

$$a > \lambda > b > \mu > c,$$

so we take

$$k^2 = \frac{a-b}{a-c}, \qquad \mathrm{sn}^2(p, k) = \frac{a-\lambda}{a-b},$$
$$k_1^2 = \frac{c^{-1}-b^{-1}}{c^{-1}-a^{-1}}, \quad \mathrm{sn}^2(q, k_1) = \frac{c^{-1}-\mu^{-1}}{c^{-1}-b^{-1}},$$

then

$$x = b^{\frac{1}{2}}\,\mathrm{sn}(p, k)\,\mathrm{dn}(q, k_1),$$
$$y = a^{\frac{1}{2}}\,\mathrm{cn}(p, k)\,\mathrm{cn}(q, k_1),$$
$$z = a^{\frac{1}{2}}\,\mathrm{dn}(p, k)\,\mathrm{sn}(q, k_1).$$

For the inner sheet

$$a > \mu > b > \lambda > c,$$

and so, in order to have real parameters p', q', we define the elliptic functions by

$$k'^2 = \frac{b-c}{a-c}, \qquad \mathrm{sn}^2(p', k') = \frac{\lambda-c}{b-c},$$
$$k_1'^2 = \frac{b^{-1}-a^{-1}}{c^{-1}-a^{-1}}, \quad \mathrm{sn}^2(q', k_1') = \frac{\mu^{-1}-a^{-1}}{b^{-1}-a^{-1}},$$

and then

$$x = c^{\frac{1}{2}}\,\mathrm{dn}(p', k')\,\mathrm{sn}(q', k_1'),$$
$$y = c^{\frac{1}{2}}\,\mathrm{cn}(p', k')\,\mathrm{cn}(q', k_1'),$$
$$z = b^{\frac{1}{2}}\,\mathrm{sn}(p', k')\,\mathrm{dn}(q', k_1').$$

* Appell et Lacour, *Fonctions Elliptiques*, p. 167; Weber, *Vierteljahrsschrift der Naturf. Ges. in Zurich* (1896), XLI, 82.

§ 64. TANGENT PLANES.

Let $OM^2 = v$ so that $v^{\frac{1}{2}}$ is the central perpendicular on the tangent plane to the Wave surface at (x, y, z). Now the tangent plane at Q is parallel to POR and the plane QOR is parallel to the tangent plane at P, whence by similar triangles

$$(\lambda - v)/v = QM^2/OM^2 = p_1^2/p^2,$$

or $$abcv^{-1}\mu^{-1}(\lambda - v)(\lambda - \mu) = (a - \lambda)(b - \lambda)(c - \lambda),$$

giving v in terms of the parameters λ and μ.

The direction cosines of OQ are those of the normal at P to the confocal λ, namely

$$p_1\xi/(a - \lambda), \quad p_1\eta/(b - \lambda), \quad p_1\zeta/(c - \lambda),$$

hence the direction cosines of OM are

$$v^{\frac{1}{2}}\lambda^{\frac{1}{2}}p_1\xi/a(a - \lambda), \quad v^{\frac{1}{2}}\lambda^{\frac{1}{2}}p_1\eta/b(b - \lambda), \quad v^{\frac{1}{2}}\lambda^{\frac{1}{2}}p_1\zeta/c(c - \lambda).$$

Those of OR are

$$p_2\xi/(a - \mu), \quad p_2\eta/(b - \mu), \quad p_2\zeta/(c - \mu),$$

and since OM' is at right angles to OR and OM, its direction cosines are

$$v^{\frac{1}{2}}\lambda^{\frac{1}{2}}p_1p_2\eta\zeta\frac{(b - c)(\lambda\mu - \overline{b + c}\mu + bc)}{b(b - \lambda)(b - \mu)c(c - \lambda)(c - \mu)} \quad \text{etc.,}$$

or $$-\frac{xv^{\frac{1}{2}}(\lambda\mu - \overline{b + c}\mu + bc)}{bc(\lambda - \mu)}.$$

The equation of the tangent plane is taken to be

$$lx + my + nz = 1,$$

and then $$l^2 + m^2 + n^2 = v^{-1},$$

and $$bc(\mu - \lambda)\,l = x(\lambda\mu - \overline{b + c}\mu + bc),$$

$$ca(\mu - \lambda)\,m = y(\lambda\mu - \overline{c + a}\mu + ca),$$

$$ab(\mu - \lambda)\,n = z(\lambda\mu - \overline{a + b}\mu + ab),$$

giving the coordinates of any tangent plane in terms of λ and μ.

The tangent plane may also be determined indirectly as follows. The intercept on the normal at any point Q of an ellipsoid between Q and the plane of symmetry perpendicular to the axis OA is OA^2/OM. After rotation through a right angle about OR we have fig. 14 in which OG is the projection of the axis OA upon the plane $Q'OM'$ and $Q'G$ is the normal at Q' to the Wave surface, and

$$GQ'.\,OM' = OA^2.$$

Complete the parallelogram $OQ'GH$: then the circle on $Q'H$ as diameter passes through M' and has its centre on OG and cuts

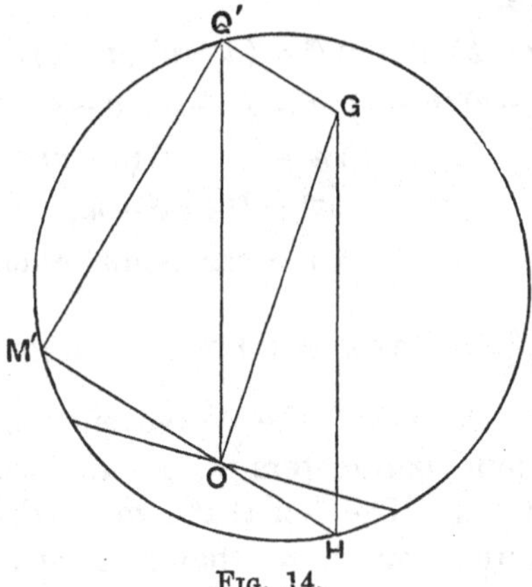

FIG. 14.

the plane $x=0$ at the ends of a chord perpendicular to OG. This chord is bisected at O and of length $2\,(M'O\,.\,OH)^{\frac{1}{2}} = 2OA$.

Hence the sphere through the circular section in the plane $x=0$ and any point of the Wave surface passes also through the projection of the centre on the tangent plane. Similarly two other spheres pass through the same two points and the circles in the other planes of reference. This theorem* gives a method of constructing the tangent plane at any point and the point of contact of a given plane.

If the equation of the tangent plane is

$$lx + my + nz = 1$$

and $v^{\frac{1}{2}}$ is the length of the central perpendicular, the foot is (lv, mv, nv). Hence the equation of the first of the three spheres is

$$lv\,(x^2 + y^2 + z^2 - a) = (v - a)\,x,$$

and this gives the relation between tangent plane and point of contact in the form

$$\frac{lv}{v-a} = \frac{x}{\lambda - a}, \quad \frac{mv}{v-b} = \frac{y}{\lambda - b}, \quad \frac{nv}{v-c} = \frac{z}{\lambda - c},$$

from which the expressions for l, m, n in terms of λ and μ may be deduced by means of the formulae already obtained. On substitution for x, y, z in the tangent plane, the tangential equation of the surface is obtained in the form

$$\frac{l^2}{v-a} + \frac{m^2}{v-b} + \frac{n^2}{v-c} = 0, \qquad (v^{-1} = l^2 + m^2 + n^2).$$

* Niven, *Quarterly Journal* (1868), IX, 22.

Let the second root of this equation, regarded as a quadratic in v, be u, then $u^{\frac{1}{2}}$ is the central perpendicular on the parallel tangent plane and we have

$$(a-b)(a-c)\,l^2 = (a-u)(a-v)\,v^{-1},$$
$$(b-c)(b-a)\,m^2 = (b-u)(b-v)\,v^{-1},$$
$$(c-a)(c-b)\,n^2 = (c-u)(c-v)\,v^{-1},$$
$$v^{-1} = l^2+m^2+n^2,$$
$$u = bcl^2+cam^2+abn^2.$$

§ 65. THE FOUR PARAMETERS.

The complete theory of the Wave surface depends on the employment of four parameters λ, μ, u, v connected by two independent relations. The fact that two apsidal surfaces remain apsidal after reciprocation shows that from any formula we may obtain another by replacing

$$x \quad y \quad z \quad a \quad b \quad c \quad \lambda \quad \mu \quad u \quad v$$

by

$$l \quad m \quad n \quad a^{-1} \quad b^{-1} \quad c^{-1} \quad v^{-1} \quad u^{-1} \quad \mu^{-1} \quad \lambda^{-1}$$

respectively. We have already proved that

$$abcv^{-1}\mu^{-1}(\lambda-v)(\lambda-\mu) = (a-\lambda)(b-\lambda)(c-\lambda),$$

whence we deduce

$$v(v-\lambda)(v-u) = (v-a)(v-b)(v-c),$$

and from these

$$(\theta-a)(\theta-b)(\theta-c)+abcv^{-1}\mu^{-1}(\theta-v)(\theta-\mu) = \theta(\theta-\lambda)(\theta-u)$$

for all values of θ. By giving θ special values, other, but not independent, results can be obtained, in particular

$$(\mu-a)(\mu-b)(\mu-c) = \mu(\mu-\lambda)(\mu-u),$$

and

$$(u-a)(u-b)(u-c) = -abcv^{-1}\mu^{-1}(u-v)(u-\mu),$$

and

$$bcv^{-1}\mu^{-1}(a-v)(a-\mu) = (a-\lambda)(a-u).$$

Further, the differentials of the four parameters are connected by the two relations included in

$$abcv^{-2}\mu^{-2}\{v(\theta-v)\,d\mu+\mu(\theta-\mu)\,dv\} = (\theta-u)\,d\lambda+(\theta-\lambda)\,du.$$

Notice how λ and u play symmetrical parts in these formulae, as also do μ and v. For the inner sheet

$$b > {\lambda \atop v} > c, \quad a > {\mu \atop u} > b;$$

and for the outer sheet

$$a > \lambda > b, \quad b > {\mu \atop u} > c, \quad a > v > ac/(a+c-b).$$

§ 66. CURVATURE.

Let L, M, N be the direction cosines of the normal at (x, y, z), then a principal radius and direction of curvature are given by

$$\rho = \frac{dx}{dL} = \frac{dy}{dM} = \frac{dz}{dN} = \frac{\Sigma x dx}{\Sigma x dL} = \frac{\Sigma a x dx}{\Sigma a x dL}.$$

Now from the formulae already obtained

$$\tfrac{1}{2}d\lambda = xdx + ydy + zdz,$$

$$-\tfrac{1}{2}abc\mu^{-2}d\mu = axdx + bydy + czdz,$$

$$v^{\frac{1}{2}} = Lx + My + Nz,$$

$$0 = Ldx + Mdy + Ndz,$$

since the direction of curvature is in the tangent plane, whence by differentiation

$$\tfrac{1}{2}v^{-\frac{1}{2}}dv = xdL + ydM + zdN.$$

It remains only to express $\Sigma a x dL$ in terms of the four parameters. Now

$$(a-b)(a-c)L^2 = (a-u)(a-v),$$

whence
$$\frac{2dL}{L} = \frac{du}{u-a} + \frac{dv}{v-a}.$$

Again it is easily proved that

$$(a-b)(a-c)Lx = v^{\frac{1}{2}}(a-\lambda)(a-u) = bc\mu^{-1}v^{-\frac{1}{2}}(a-\mu)(a-v).$$

Hence
$$2\Sigma a x dL = 2\Sigma a(Lx)\,dL/L$$

$$= \Sigma a v^{\frac{1}{2}} \frac{\lambda - a \,.\, u - a}{a - b \,.\, a - c} \cdot \frac{du}{u-a}$$

$$+ \Sigma a v^{\frac{1}{2}} \frac{bc}{\mu v} \cdot \frac{\mu - a \,.\, v - a}{a - b \,.\, a - c} \cdot \frac{dv}{v-a}$$

$$= -v^{\frac{1}{2}}du.$$

Hence*
$$\rho = \frac{v^{\frac{1}{2}}d\lambda}{dv} = \frac{abc\mu^{-2}d\mu}{v^{\frac{1}{2}}du},$$

giving the differential equation of the lines of curvature in the form

$$v\mu^2 d\lambda du = abc\, d\mu dv.$$

When expressed in terms of λ and v only this takes the form

$$d\lambda^2 - \left(\frac{\lambda-a}{v-a} + \frac{\lambda-b}{v-b} + \frac{\lambda-c}{v-c} - \frac{\lambda}{v}\right)d\lambda dv$$

$$+ \frac{(\lambda-a)(\lambda-b)(\lambda-c)}{(v-a)(v-b)(v-c)}dv^2 = 0,$$

* The first expression for ρ is the analogue of the formula $r\,dr/dp$ for plane curves.

and since the first equation is symmetrical in λ and u, it follows that λ may be replaced by u in the last equation. The equation giving the radii of curvature is obtained by substituting

$$d\lambda/dv = v^{\frac{1}{2}}\rho.$$

The coefficients have geometrical interpretations, for if N_1 is the intercept on the normal by the plane $x=0$,

$$N_1 = x/lv^{\frac{1}{2}} = v^{\frac{1}{2}}(\lambda - a)/(v-a).$$

Hence $$\rho^2 - (N_1 + N_2 + N_3 - \lambda v^{-\frac{1}{2}})\rho + N_1 N_2 N_3 v^{-\frac{1}{2}} = 0.$$

If $$f(\lambda) \equiv (\lambda - a)(\lambda - b)(\lambda - c),$$

$$f'(\lambda)/f(\lambda) = (\lambda - a)^{-1} + (\lambda - b)^{-1} + (\lambda - c)^{-1},$$

the differential equation of lines of curvature can be written in the form

$$d\lambda^2 - \{(\lambda - v) f'(v)/f(v) + 3 - \lambda/v\}\, d\lambda dv + dv^2 f(\lambda)/f(v) = 0.$$

This has been integrated when f is quadratic but not when, as in the general case, f is cubic*.

$$\frac{(N_1 - v^{\frac{1}{2}})(N_2 - N_3)}{b-c} = \frac{(N_2 - v^{\frac{1}{2}})(N_3 - N_1)}{c-a} = \frac{(N_3 - v^{\frac{1}{2}})(N_1 - N_2)}{a-b},$$

showing that the intersections of the normal with the planes of reference and perpendicular central plane form a range of constant cross ratio.

The line element is given by

$$4ds^2 = \frac{d\lambda^2}{\lambda - v} + \frac{abcd\mu^2}{\mu^3(u-\mu)}.$$

§ 67. ASYMPTOTIC LINES.

The differential equation of asymptotic lines is

$$dx\,dl + dy\,dm + dz\,dn = 0,$$

or $$\Sigma lx\left(\frac{d\lambda}{\lambda - a} + \frac{a\,d\mu}{u(\mu - a)}\right)\left(\frac{du}{u-a} + \frac{dv}{v-a}\right) = 0;$$

further $$lx = \frac{(\lambda - a)(u-a)}{(a-b)(a-c)} = \frac{bc}{\mu v}\frac{(\mu - a)(v-a)}{(a-b)(a-c)}.$$

The coefficients of $d\lambda du$ and $d\mu dv$ are 0, and the equation reduces to

$$\mu(\mu - u)\, d\lambda dv + v(v - \lambda)\, d\mu du = 0.$$

This equation is unaltered if λ, μ, u, v are replaced by

$$v^{-1}, u^{-1}, \mu^{-1}, \lambda^{-1},$$

respectively, illustrating the fact that asymptotic lines are reciprocated into asymptotic lines.

* Darboux, *Comptes Rendus*, XCII, 446; XCVII, 1133.

When expressed in terms of u and λ only, this equation takes the form

$$\frac{du^2}{(u-a)(u-b)(u-c)}=\frac{d\lambda^2}{(\lambda-a)(\lambda-b)(\lambda-c)}.$$

To integrate we use the theorem that if three points (x_1y_1), (x_2y_2), (x_3y_3), of the curve

$$y^2=f(x)\equiv(x-a)(x-b)(x-c)$$

are collinear, then

$$dx_1/y_1+dx_2/y_2+dx_3/y_3=0.$$

Hence the equation

$$du/\sqrt{f(u)}+d\lambda/\sqrt{f(\lambda)}=0$$

expresses that the points $(u, \sqrt{f(u)})$ and $(\lambda, \sqrt{f(\lambda)})$ are collinear with a fixed point (x_1y_1) on the curve. If the equation of their line is

$$y=mx+n,$$

then

$$(x-a)(x-b)(x-c)-(mx+n)^2\equiv(x-x_1)(x-u)(x-\lambda);$$

and

$$(ma+n)^2=(x_1-a)(a-u)(a-\lambda).$$

The required integral is obtained by eliminating m and n; x_1 is the constant of integration: instead of it we introduce

$$\alpha=(b-c)(x_1-a),$$
$$\beta=(c-a)(x_1-b),$$
$$\gamma=(a-b)(x_1-c),$$

so that

$$\alpha+\beta+\gamma=0,$$

and then the integral is*

$$\alpha^{\frac{1}{2}}\sqrt{\frac{(a-u)(a-\lambda)}{(a-b)(a-c)}}+\beta^{\frac{1}{2}}\sqrt{\frac{(b-u)(b-\lambda)}{(b-c)(b-a)}}+\gamma^{\frac{1}{2}}\sqrt{\frac{(c-u)(c-\lambda)}{(c-a)(c-b)}}=0.$$

This is the same as

$$\sqrt{\alpha lx}+\sqrt{\beta my}+\sqrt{\gamma nz}=0,$$

which, again, is the point-plane equation of a tetrahedral complex (p. 69). The inference is that at each point of an asymptotic curve the complex cone of a certain tetrahedral complex *touches* the surface. By varying the constant of the complex all the asymptotic lines are obtained.

In terms of μ and v the equation is

$$\frac{\mu^{-3}d\mu^2}{(a-\mu)(b-\mu)(c-\mu)}=\frac{v^{-3}dv^2}{(a-v)(b-v)(c-v)}.$$

* Darboux, *Théorie des Surfaces*, I, 143.

As in the general Kummer surface, the asymptotic curves have an envelope consisting of the singular conics, and a cusp locus which reduces to isolated nodes. Hence the *elliptic* and *hyperbolic* regions of the Wave surface are separated by the circles of contact of the four tropes and the four nodes. It is easy to see on which side of these boundaries the curves lie and that the hyperbolic regions consist of four detached portions, each bounded by one circle and one node (fig. 13, p. 104). A complete asymptotic curve consists of four branches, one in each portion, and each branch touches the circle at one point and has a cusp at the node.

There are two elliptic regions, namely the entire inner sheet, and the outer sheet bounded by the four circles.

§ 68. PAINVIN'S COMPLEX.

The quadratic complex of which the wave surface is the singular surface is Painvin's complex* of lines through which the tangent planes to a quadric are at right angles (cf. the generation of the harmonic complex, p. 97, § 58).

Let the quadric be

$$x^2/\alpha + y^2/\beta + z^2/\gamma = 1,$$

then the complex, with current line coordinates l, m, n, l', m', n', is

$$(\beta+\gamma)\, l^2 + (\gamma+\alpha)\, m^2 + (\alpha+\beta)\, n^2 = l'^2 + m'^2 + n'^2.$$

The complex cone of rays through any point is the director cone of the enveloping cone of tangents to the quadric: the latter referred to its own axes has equation

$$x^2/\lambda + y^2/\mu + z^2/\nu = 0,$$

where λ, μ, ν are the parameters of the confocals through the point. Hence the complex cone is

$$(\mu+\nu)\, x^2 + (\nu+\lambda)\, y^2 + (\lambda+\mu)\, z^2 = 0$$

referred to its axes. For a singular point this must break into two planes, which can happen only if

$$(\mu+\nu)(\nu+\lambda)(\lambda+\mu) = 0.$$

Now if (x, y, z) is the singular point, λ, μ, ν are the roots of the equation in λ,

$$\frac{x^2}{\alpha-\lambda} + \frac{y^2}{\beta-\lambda} + \frac{z^2}{\gamma-\lambda} = 1,$$

whence $$\lambda + \mu + \nu = \alpha + \beta + \gamma - x^2 - y^2 - z^2.$$

* Painvin, *Nouvelles Annales* (1872), II, 49; Sturm, *Liniengeometrie*, III, 35.

The condition for a singular point shows that $\lambda + \mu + \nu$ is a root and hence the equation of the singular surface is

$$\frac{x^2}{x^2+y^2+z^2-\beta-\gamma}+\frac{y^2}{x^2+y^2+z^2-\gamma-\alpha}+\frac{z^2}{x^2+y^2+z^2-\alpha-\beta}=1.$$

In order to make this agree with the former notation for the wave surface we must put

$$a=\beta+\gamma, \quad b=\gamma+\alpha, \quad c=\alpha+\beta.$$

Thus we find that the wave surface may be generated by lines of curvature on confocal quadrics, being the intersections of confocals whose parameters are equal and opposite. Taking $\lambda+\nu=0$ we have the parametric expression of the wave surface in the form

$$(\alpha-\beta)(\alpha-\gamma)x^2=(\alpha^2-\lambda^2)(\alpha-\mu),$$
$$(\beta-\gamma)(\beta-\alpha)y^2=(\beta^2-\lambda^2)(\beta-\mu),$$
$$(\gamma-\alpha)(\gamma-\beta)z^2=(\gamma^2-\lambda^2)(\gamma-\mu).$$

It is easy to see that the curves $\mu=\text{const.}$, $\lambda=\text{const.}$, are the same as those which in the former notation were $\lambda=\text{const.}$, $\mu=\text{const.}$

The singular line at a singular point is the line of intersection of the two singular planes

$$(\mu-\lambda)x^2+(\lambda+\mu)z^2=0$$

referred to the three normals to confocals. Hence it is the normal to the confocal μ and tangent to the line $\lambda=\text{const.}$

These two planes touch a sphere whose centre is at the origin and whose radius is independent of μ*.

Having considered the rays through any point we next consider the rays lying in any plane π. Let any ray cut the section of the director sphere by π in P and Q. Then the planes through P and Q perpendicular to PQ must touch the ellipsoid. It is easy to deduce from this that PQ touches a conic confocal with the projection of the ellipsoid on the plane π.

Hence as a plane moves parallel to itself the complex curves are confocal conics. It is a singular plane when it touches the wave surface and then the rays in it pass through the foci. When the plane touches the outer sheet it contains one real ray, the minor axis, which is singular. When the plane touches the inner sheet the major axis is the singular ray and cuts the outer sheet at the foci.

* Böklen, *Zeitschrift für Math.* XXVII, 160.

Prove that the plane joining PQ to the pole of π touches a confocal whose parameter is minus that of the confocal touched by π.

The quadratic complex being

$$a^2l^2+b^2m^2+c^2n^2=l'^2+m'^2+n'^2$$

we see that the fundamental complexes are given by

$$\left.\begin{aligned} a^{\frac{1}{2}}x_1&=al+il'\\ ia^{\frac{1}{2}}x_2&=al-il'\end{aligned}\right\} \quad \left.\begin{aligned} b^{\frac{1}{2}}x_3&=bm+im'\\ b^{\frac{1}{2}}x_4&=bm-im'\end{aligned}\right\} \quad \left.\begin{aligned} c^{\frac{1}{2}}x_5&=cn+in'\\ c^{\frac{1}{2}}x_6&=cn-in'\end{aligned}\right\}$$

and are pairs of conjugate imaginaries. The coefficients k_s in the complex are also conjugate in pairs if we take

$$k_1=-k_2=ia,\quad k_3=-k_4=ib,\quad k_5=-k_6=ic.$$

The coordinates of any tangent line of the Wave surface are given (p. 58) by

$$\rho x_s^2=(k_s-\lambda)^2\,(k_s-\mu_1)\,(k_s-\mu_2)/f'\,(k_s)$$

where μ_1 and μ_2 are the parameters of the asymptotic lines through the point of contact; this gives

$$\rho'\,(al+il')=(ia-\lambda)\sqrt{(b^2-c^2)\,(ia-\mu_1)\,(ia-\mu_2)},$$

etc.

From the relations connecting line coordinates (l, m, n, l', m', n') with point coordinates, it is easy to deduce

$$\Sigma\,(a^2-\nu^2)\,(b^2-c^2)\,(a^2+\mu_1{}^2)^{\frac{1}{2}}\,(a^2+\mu_2{}^2)^{\frac{1}{2}}+(b^2-c^2)\,(c^2-a^2)\,(a^2-b^2)=0$$

and $$\Sigma\,(b^2c^2-\eta)\,(b^2-c^2)\,(a^2+\mu_1{}^2)^{\frac{1}{2}}\,(a^2+\mu_2{}^2)^{\frac{1}{2}}+\mu_1\mu_2(\quad)(\quad)(\quad)=0,$$

where

$$\nu^2=x^2+y^2+z^2,$$
$$\eta=a^2x^2+b^2y^2+c^2z^2,$$

giving the connection between λ, μ, μ_1, μ_2.

It is easy to verify them as integrals of the equation of asymptotic lines.

CHAPTER XI.

REALITY AND TOPOLOGY.

§ 69. REALITY OF THE COMPLEXES.

In this chapter we examine and distinguish the different kinds of Kummer's surface which have sixteen *distinct* nodes and are given by real equations, the surface itself not being necessarily real*. We are not here concerned with degenerate cases in which the nodes coincide†.

The equation is completely determined by six apolar complexes and one arbitrary node. Since the equation is to be real, all imaginaries must occur in pairs and the primary classification is according to the number of complexes which are real. We arrange them in three pairs 12, 34, 56 and consider in turn the cases when three, two, one, or none of these pairs are real. In all cases three congruences are real and their directrices form a real fundamental tetrahedron 12 . 34 . 56 which will be taken for reference. Then by taking suitable real multiples of the point coordinates it can be arranged that in the four cases respectively

$$\text{I.}\quad \begin{aligned} x_1 &= p_{14} - p_{23}, & x_3 &= p_{24} - p_{31}, & x_5 &= p_{34} - p_{12},\\ ix_2 &= p_{14} + p_{23}, & ix_4 &= p_{24} + p_{31}, & ix_6 &= p_{34} + p_{12}, \end{aligned}$$

$$\text{II.}\quad \begin{aligned} x_1 &= i^{\frac{1}{2}}p_{14} - i^{-\frac{1}{2}}p_{23}, & x_3 &= p_{24} - p_{31}, & x_5 &= p_{34} - p_{12},\\ ix_2 &= i^{\frac{1}{2}}p_{14} + i^{-\frac{1}{2}}p_{23}, & ix_4 &= p_{24} + p_{31}, & ix_6 &= p_{34} + p_{12}, \end{aligned}$$

$$\text{III.}\quad \begin{aligned} x_1 &= p_{14} - p_{23}, & x_3 &= i^{\frac{1}{2}}p_{24} - i^{-\frac{1}{2}}p_{31}, & x_5 &= i^{\frac{1}{2}}p_{34} - i^{-\frac{1}{2}}p_{12},\\ ix_2 &= p_{14} + p_{23}, & ix_4 &= i^{\frac{1}{2}}p_{24} + i^{-\frac{1}{2}}p_{31}, & ix_6 &= i^{\frac{1}{2}}p_{34} + i^{-\frac{1}{2}}p_{12}, \end{aligned}$$

$$\text{IV.}\quad \begin{aligned} x_1 &= i^{\frac{1}{2}}p_{14} - i^{-\frac{1}{2}}p_{23}, & x_3 &= i^{\frac{1}{2}}p_{24} - i^{-\frac{1}{2}}p_{31}, & x_5 &= i^{\frac{1}{2}}p_{34} - i^{-\frac{1}{2}}p_{12},\\ ix_2 &= i^{\frac{1}{2}}p_{14} + i^{-\frac{1}{2}}p_{23}, & ix_4 &= i^{\frac{1}{2}}p_{24} + i^{-\frac{1}{2}}p_{31}, & ix_6 &= i^{\frac{1}{2}}p_{34} + i^{-\frac{1}{2}}p_{12}, \end{aligned}$$

* Rohn, *Math. Ann.* XVIII, 99, may be consulted for a more detailed account.

† See an exhaustive paper by Weiler, *Math. Ann.* VII, 145.

so that in all cases $\Sigma x_s^2 = -4\Sigma p_{14} p_{23} = 0$. When the multiples of the point coordinates are not restricted to be real, all the cases can be reduced to the first by the substitutions

$$
\begin{array}{lllll}
\text{I.} & x, & y, & z, & t, \\
\text{II.} & x, & i^{\frac{1}{2}}y, & i^{\frac{1}{2}}z, & t, \\
\text{III.} & ix, & i^{\frac{1}{2}}y, & i^{\frac{1}{2}}z, & t, \\
\text{IV.} & x, & y, & z, & it.
\end{array}
$$

The following table shows which of the directrices, tetrahedra, and quadrics are real in the four cases.

Case	I	II	III	IV
Tetrahedra	12.34.56 12.36.54 14.36.52 14.32.56 16.32.54 16.34.52	12.34.56 12.36.54	12.34.56	12.34.56
Quadrics	all except 135	123 124 125 126	134 234	135 235 145 245

The surface is the singular surface of a family of quadratic complexes

$$(k_1 - \lambda)^{-1}x_1^2 + (k_2 - \lambda)^{-1}x_2^2 + (k_3 - \lambda)^{-1}x_3^2 + (k_4 - \lambda)^{-1}x_4^2 + (k_5 - \lambda)^{-1}x_5^2 + (k_6 - \lambda)^{-1}x_6^2 = 0;$$

it is not necessary that this complex should be real for even one value of λ, but only that the conjugate imaginary complex should be included in the family, for then the singular surface, having the same relation to two conjugate imaginary complexes, must have a real equation.

Let μ be the parameter of the latter complex, then by means of the relation $\Sigma x_s^2 = 0$ the equations of the two complexes can be written

$$\Sigma \frac{k_s - \mu}{k_s - \lambda} x_s^2 = 0 \quad \text{and} \quad \Sigma \frac{k_s - \lambda}{k_s - \mu} x_s^2 = 0.$$

Since the singular surface is unaltered when the same linear transformation is performed on all the k_s, we shall replace

$$(k_s - \mu)(k_s - \lambda)^{-1}$$

by k_s, and then a sufficient condition that the equation of the singular surface may be real is that the complexes

$$\Sigma k_s x_s^2 = 0, \quad \Sigma k_s^{-1} x_s^2 = 0$$

be conjugate imaginaries. Put $k_s = r_s e^{i\theta_s}$; then in case I we must have

$$r_s e^{-i\theta_s} = \rho r_s^{-1} e^{-i\theta_s} \qquad (s = 1, 2, 3, 4, 5, 6)$$

where ρ is a factor of proportionality, whence

$$r_1^2 = r_2^2 = r_3^2 = r_4^2 = r_5^2 = r_6^2 = \rho,$$

showing that ρ is real and positive, and that the points representing k_s in the plane of the complex variable $re^{i\theta}$ lie on a circle.

In case II x_1 and x_2 are conjugate and we must have

$$r_1 e^{-i\theta_1} = \rho r_2^{-1} e^{-i\theta_2},$$
$$r_2 e^{-i\theta_2} = \rho r_1^{-1} e^{-i\theta_1},$$
$$r_s e^{-i\theta_s} = \rho r_s^{-1} e^{-i\theta_s} \qquad (s = 3, 4, 5, 6)$$

whence
$$r_1 r_2 = r_3^2 = r_4^2 = r_5^2 = r_6^2 = \rho$$

and
$$\theta_1 = \theta_2,$$

showing that the points $k_1 k_2$ are inverse with respect to the circle on which $k_3 k_4 k_5 k_6$ lie.

Similarly in case III

$$r_1^2 = r_2^2 = r_3 r_4 = r_5 r_6 = \rho,$$

$$\theta_3 = \theta_4, \quad \theta_5 = \theta_6,$$

showing that $k_3 k_4$ and $k_5 k_6$ are two pairs of inverse points with respect to a circle on which $k_1 k_2$ lie. In these three cases ρ is necessarily positive, but in case IV ρ may be either positive or negative; if ρ is positive

$$r_1 r_2 = r_3 r_4 = r_5 r_6 = \rho,$$

$$\theta_1 = \theta_2, \quad \theta_3 = \theta_4, \quad \theta_5 = \theta_6$$

and the six points k_s form three pairs of inverse points; if ρ is negative

$$r_1 r_2 = r_3 r_4 = r_5 r_6 = -\rho,$$

$$\theta_1 - \theta_2 = \theta_3 - \theta_4 = \theta_5 - \theta_6 = \pi$$

and the point k_{2s-1} is obtained from k_{2s} by an inversion followed by a reflexion in the origin.

When ρ is positive it is convenient to effect a linear substitution on the complex variable so as to transform the circle into the real axis; then in cases I, II, III and the first subcase of IV *the coefficients* k_s *may be taken to be either real or pairs of conjugate imaginaries* according to the reality of the corresponding complexes. In the second subcase of IV no further simplification is possible.

§ 70. SIX REAL FUNDAMENTAL COMPLEXES.

The different kinds of surface included in case I are distinguished by the order of magnitude of the six coefficients k_s. These quantities are the parameters of the six nodes on any singular conic, and their order of magnitude determines the cyclical order in which the nodes follow each other consecutively.

The order of k_1, k_3, k_5 among themselves is immaterial, since a permutation of odd suffixes is equivalent to the choice of a new real fundamental tetrahedron of reference; similarly for k_2, k_4, k_6, so that the only permutations which are essentially distinct are 123456, 123465, 126453.

I *a*. *Sixteen real nodes.*

We begin by investigating the shape of the surface when

$$k_1 > k_2 > k_3 > k_4 > k_5 > k_6.$$

The pencil of tangents at any point (μ_1, μ_2) of the surface is given (p. 58) by

$$x_s^2 f'(k_s) = c(k_s - \mu_1)(k_s - \mu_2)(k_s - \lambda)^2$$

where

$$f(\mu) \equiv (\mu - k_1)(\mu - k_2)(\mu - k_3)(\mu - k_4)(\mu - k_5)(\mu - k_6)$$

and c is a factor of proportionality.

For a real line x_s^2 is alternately positive and negative and so also is $f'(k_s)$ for $s = 1, 2, 3, 4, 5, 6$ so that all the left sides have the same sign. Hence μ_1 and μ_2 are either conjugate imaginaries, or are real and lie between the same two consecutive k_s. In the former case the inflexional tangents $\lambda = \mu_1$ and $\lambda = \mu_2$ are imaginary and the point is *elliptic*; in the latter case the inflexional tangents are real and the point is *hyperbolic*. The boundaries of elliptic and hyperbolic regions are given by $\mu_1 = \mu_2$, that is by the singular conics and nodes. Now if (x) is a real line, so also all the lines obtained from it by changing the signs of the coordinates are real; since it is possible to have $\mu_1 = \mu_2$ it follows that *there are sixteen real nodes and sixteen real tropes.*

The region described by a point whose parameters μ_1 and μ_2 take all real values between k_s and k_{s+1} is a connected portion of the surface bounded by two arcs of conics terminated by nodes whose parameters on either of them are k_s and k_{s+1} (for the same collineation which interchanges two nodes interchanges also the two conics through them both). Each pair of consecutive coefficients k_s, k_{s+1} gives eight hyperbolic segments and hence there are *forty-eight* in all; of the thirty-two points having the same parameters μ_1, μ_2, four lie on each of eight segments.

Consider the course of any one asymptotic curve $\mu_1 = \text{const.}$ over one segment. It has a cusp at each node and touches each conic once at points given by $\mu_2 = \mu_1$. The extreme values of μ_2 give the points where the curve crosses the two principal asymptotic curves $\mu = k_s$ and $\mu = k_{s+1}$; at each of these points the curve has an apparent inflexion because the tangent to it has four-point contact with the surface. The figure shows two consecutive nodes on two conics with the three points whose parameters are k_s, μ, k_{s+1} indicated on each of them. As μ varies from k_s to k_{s+1} the asymptotic curve μ sweeps out the whole segment twice and the two branches of the curve coincide in the case of the principal asymptotic curves. The latter meet in a point which is on one of the directrices of the congruence $x_s = 0$, $x_{s+1} = 0$, and is a point where both inflexional tangents have four-point contact.

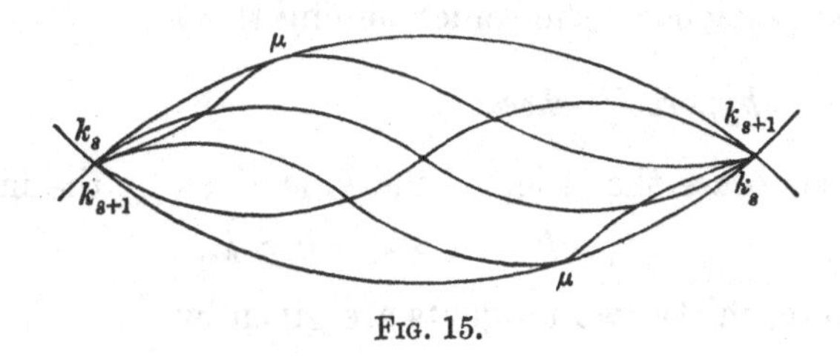

Fig. 15.

The elliptic region of the surface is given by

$$\mu_1 = \mu + i\mu', \quad \mu_2 = \mu - i\mu',$$

μ and μ' being real, and the complete boundary by

$$\mu' = 0, \quad -\infty < \mu < +\infty.$$

For any value of μ the line given by

$$x_s \sqrt{f'(k_s)} = \sqrt{c}\,(k_s - \mu)(k_s - \lambda)$$

either lies in a trope or passes through a node, and as μ passes through any one of the six values k_s, one coordinate changes sign and accordingly the line passes from one state to the other. Hence as μ takes all possible values the variable point of contact describes three arcs of conics joining three nodes and forming the boundary of an *elliptic triangle*; at the same time the points of contact of the lines obtained from (x) by changing the signs of the coordinates describe the boundaries of other elliptic triangles so that the number of them is *thirty-two*. The two nodes on each side of any one elliptic triangle are interchanged either by one of the collineations 12, 34, 56 or else by one of 23, 45, 61. Hence the nodes of the Göpel tetrad 0, 12, 34, 56 are the corners of a tetrahedral portion of the surface (p. 22) having four elliptic triangles for faces and six hyperbolic segments for edges. Each pair of opposite hyperbolic segments is cut by an edge of the fundamental tetrahedron 12.34.56 of which one corner is surrounded by the tetrahedral piece. The remaining three tetrads of the group-set containing 0, 12, 34, 56 give other tetrahedral pieces surrounding the other three corners of the tetrahedron 12.34.56, and divided into elliptic and hyperbolic regions in a similar way. Again, four other tetrahedral pieces have for corners the Göpel tetrads of the group-set containing 0, 23, 45, 61.

Thus the whole surface consists of eight pieces containing forty-eight hyperbolic segments and thirty-two elliptic triangles; each piece is attached at its nodes to each of four other pieces. This is clearly shown in the frontispiece where thirteen nodes are actually visible and portions of the conics joining them.

I *b*. *Two hyperbolic sheets.*

In this subcase the order of the k_s is given by the inequalities

$$k_1 > k_2 > k_3 > k_4 > k_6 > k_5.$$

As before, thirty-two tangents are given by

$$x_s^2 f'(k_s) = c\,(k_s - \mu_1)(k_s - \mu_2)(k_s - \lambda)^2$$

but now the signs of the left sides are $+, +, +, +, -, -$, when the lines are real, so that μ_1 and μ_2 cannot be imaginary, and the surface has everywhere negative curvature.

The real values μ_1, μ_2 must satisfy the inequalities implied by the cyclical order

$$\mu_1,\ k_1,\ k_2,\ k_3,\ k_4,\ \mu_2,\ k_6,\ k_5,\ \mu_1$$

so that it is impossible to have $\mu_1 = \mu_2$, and *all the nodes and tropes are imaginary.*

There are only four real principal asymptotic curves, namely those whose parameters are k_1, k_4, k_6, k_5. As μ_2 varies from k_4 to k_6 the corresponding asymptotic curve sweeps out a connected region of the surface, which, since the envelope of asymptotic curves is imaginary, must be a sheet without a boundary like a hyperboloid of one sheet. Along any one curve μ_1 varies from k_5 through infinity to k_1, giving the intersections with another family of asymptotic curves covering the same sheet. By continuously varying

μ_1 and μ_2 it is possible to change the signs of all the coordinates of a tangent line except x_2 and x_3. Hence of thirty-two associated points sixteen lie on one sheet, and we conclude that *the whole surface consists of two infinite hyperbolic sheets.* These are cut in real points by the directrices (14), (16), (54), (56) and the corresponding collineations transform each sheet into itself. The remaining five real pairs of directrices do not meet the surface in real points; and of the corresponding collineations, (12), (52), (43), (63) change the sign of x_2x_3 and hence interchange the sheets, while (23) transforms each sheet into itself. Any ray of the congruence (23) cuts each sheet in two points forming a harmonic range with the points where the ray meets the directrices. Hence *these directrices are separated by both sheets,* just as a line is separated from its polar with respect to a hyperboloid of one sheet which does not cut it.

I *c*. *Imaginary surface.*

In the third subcase

$$k_1 > k_2 > k_6 > k_4 > k_5 > k_3$$

and the signs of $x_s^2 f'(k_s)$ are +, +, −, +, +, −, for $s = 1, 2, 3, 4, 5, 6$ respectively showing that μ_1 and μ_2 must be separated by k_6 and k_3 and by no other of the k_s, which is impossible. Hence the surface has no real tangents and is therefore altogether imaginary.

§ 71. EQUATIONS OF SURFACES I *a*, I *b*, I *c*.

The equation of Kummer's surface with sixteen nodes $(\alpha, \beta, \gamma, \delta)$ etc. referred to a fundamental tetrahedron is (pp. 81, 82)

$$x^4 + y^4 + z^4 + t^4 + A(x^2t^2 + y^2z^2) + B(y^2t^2 + z^2x^2) + C(z^2t^2 + x^2y^2) + Dxyzt = 0,$$

where $$2 - A = \frac{(\alpha^2+\delta^2)^2 - (\beta^2+\gamma^2)^2}{\alpha^2\delta^2 - \beta^2\gamma^2} = 4\frac{(k_3-k_5)(k_6-k_4)}{(k_3-k_4)(k_6-k_5)},$$

$$2 - B = \frac{(\beta^2+\delta^2)^2 - (\gamma^2+\alpha^2)^2}{\beta^2\delta^2 - \gamma^2\alpha^2} = 4\frac{(k_5-k_1)(k_2-k_6)}{(k_5-k_6)(k_2-k_1)},$$

$$2 - C = \frac{(\gamma^2+\delta^2)^2 - (\alpha^2+\beta^2)^2}{\gamma^2\delta^2 - \alpha^2\beta^2} = 4\frac{(k_1-k_3)(k_4-k_2)}{(k_1-k_2)(k_4-k_3)},$$

and $$D = \alpha\beta\gamma\delta(2-A)(2-B)(2-C)(\delta^2+\alpha^2+\beta^2+\gamma^2)^{-2}$$

$$= 4\frac{k_1k_2(k_3+k_4-k_5-k_6) + k_3k_4(k_5+k_6-k_1-k_2) + k_5k_6(k_1+k_2-k_3-k_4)}{(k_1-k_2)(k_3-k_4)(k_5-k_6)}.$$

If $\alpha, \beta, \gamma, \delta$ are real, all the nodes are real, and we have the equation I a. If any of the ratios $\alpha : \beta : \gamma : \delta$ are imaginary, all the nodes are imaginary and the equation represents I b or I c. To obtain I b we have simply to interchange k_5 and k_6 in the coefficients, which has the effect of changing the signs of A, B and D; accordingly the equation I b is

$$x^4 + y^4 + z^4 + t^4 - A(x^2t^2 + y^2z^2) - B(y^2t^2 + z^2x^2) + C(z^2t^2 + x^2y^2) - Dxyzt = 0.$$

This can also be obtained from Ia either by replacing $\alpha, \beta, \gamma, \delta$ by $i\alpha, i\beta, \gamma, \delta$ respectively or by replacing x, y, z, t by ix, iy, z, t, and it is then obvious that all the nodes are imaginary. The equation I c can be obtained from I a by interchanging k_3 and k_6, and from I b by the real linear transformation which interchanges x_3 and x_5. But it is interesting to notice that the preceding equation after the coefficients are expressed in terms of $\alpha, \beta, \gamma, \delta$ is typical of both I b and I c according to the sign of $2 - C$. For to every real point (x, y, z, t) on I b correspond a pair of imaginary points (ix, iy, z, t) and $(-ix, -iy, z, t)$ on I a lying on a real ray of the congruence (56). Now certain rays of this congruence cut I a in four real points (for example lines joining two nodes), and the passage to imaginary intersections can take place only through positions in which two coincide, that is, intersections of the surface with the directrices (56). Now the line $x = 0$, $y = 0$ cuts I a where

$$z^4 + t^4 + Cz^2t^2 = 0$$

and all the intersections are real if $C < 2$ and all are imaginary if $C > 2$, so that these inequalities are necessary and sufficient to distinguish I b and I c.

§ 72. FOUR REAL AND TWO IMAGINARY COMPLEXES.

In case II k_1 and k_2 are conjugate imaginaries and the shape depends upon the cyclical order of k_3, k_4, k_5, k_6 which may be of two essentially different kinds, according as the odd suffixes are separated by the even suffixes or not.

II a. *Eight real nodes**.

Consider first the case when

$$k_3 > k_4 > k_5 > k_6.$$

* A figure of this surface is given in the *Catalog mathematischer Modelle*, Halle (1903), p. 92, of Martin Schilling, from whom models of Kummer surfaces with sixteen, eight, and four real nodes may be obtained.

Any tangent line is given by

$$x_s^2 f'(k_s) = c(k_s - \mu_1)(k_s - \mu_2)(k_s - \lambda)^2,$$

and for a real line $x_1^2 f'(k_1)$ and $x_2^2 f'(k_2)$ are conjugate imaginaries while, for $s = 3, 4, 5, 6$, $x_s^2 f'(k_s)$ is positive. Hence μ_1 and μ_2 may be real or conjugate imaginaries; if real, they must lie between the same pair of consecutive k_s taken in cyclical order. The hyperbolic segments are of the same nature as in I a, but their number is only thirty-two and there are only four real principal asymptotic curves $\mu = k_3$, k_4, k_5 or k_6.

Since for a real line x_1 and x_2 are conjugate imaginaries, only sixteen out of a group of thirty-two lines are real. Hence there are only eight real nodes and eight real tropes, namely 0, 12, 34, 56, 35, 36, 45, 46.

At points of an elliptic segment, μ_1 and μ_2 have the form $\mu \pm i\mu'$, and the complete boundary is described when $\mu' = 0$ and μ takes all real values. As μ varies the point of contact (μ, μ) of the pencil of tangents $(\mu, \mu, \lambda, \lambda)$ describes an arc of a conic and remains at a node, alternately; after μ has made the complete cycle of real values x_1 and x_2 recover their original values but x_3, x_4, x_5, x_6 have all changed their signs. The pencil of tangents returns to its original position only after μ has taken every real value *twice* and then the point of contact has described four arcs and passed through four nodes. Hence the elliptic segments have four sides and four corners each.

The collineations which interchange the nodes at the ends of a side are either 34 and 56 or 45 and 63. Hence the nodes 0, 34, 12, 56 taken in this order are the corners of two elliptic segments, being joined consecutively by hyperbolic segments; so also are the nodes of the one other real tetrad of the same group-set, namely 35, 54, 46, 36. Similarly the tetrads of nodes 0, 45, 12, 63 and 35, 43, 46, 65 are connected by elliptic segments. Each tetrad forms a four-cornered piece of the surface having two elliptic faces and four hyperbolic edges. Each of one group-set is attached at two nodes to each of the other group-set.

Consider the degenerate case when one node 0 is taken on the quadric 123 . 456. Of the harmonic conjugates of the real generators through 0 with respect to the six pairs of directrices lying on the surface and named after them, only (12) of one system and (45) (56) (64) of the other system are real, since the correlations associated with $x_1 = 0$, and $x_2 = 0$, separately, transform a real line into an imaginary one. Hence the nodes in this degenerate case lie at the intersections of two and four generators, and this gives a diagram of their relative situation in the general case.

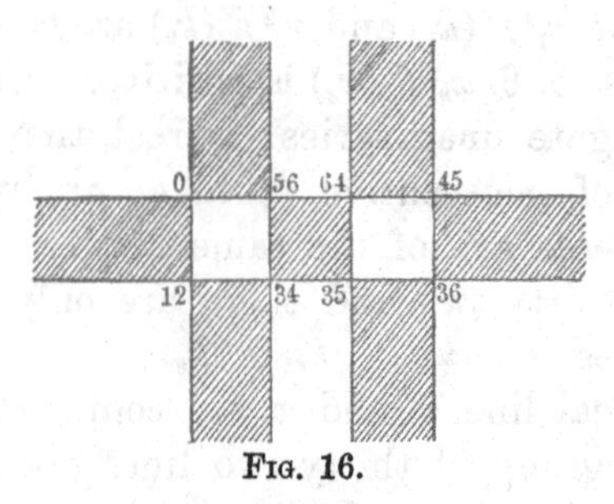

FIG. 16.

As in case I each piece is cut in pairs of points by three directrices; there is this difference to I a that each elliptic segment contains a point where the two inflexional tangents have four-point contact, but are imaginary. These points lie by fours on the directrices (12).

II b. *One hyperbolic sheet.*

Consider next the case when

$$k_3 > k_4 > k_6 > k_5.$$

The tangents $(\mu_1, \mu_2, \lambda, \lambda)$ are real only if the signs of $(k_s - \mu_1)(k_s - \mu_2)$ are $+, +, -, -$, or the opposite, for $s = 3, 4, 5, 6$ respectively. The cyclical order of magnitude must therefore be

$$\mu_1 \; k_3 \; k_4 \; \mu_2 \; k_6 \; k_5 \; \mu_1,$$

and it is impossible for μ_1 and μ_2 to be equal, or to be conjugate imaginaries, whence *there are no real nodes.* The surface is real, and everywhere hyperbolic; there are four real principal asymptotic curves, and the discussion is the same as in I b except that there is only one real sheet; for of thirty-two associated points only sixteen are real, corresponding to the different signs of x_3, x_4, x_5, x_6, and these can be reached by continuous variation of the elliptic coordinates μ_1, μ_2, and accordingly lie on the same sheet.

Equations of II a *and* II b.

The case when two fundamental complexes are imaginary can be obtained from the case when all are real by the substitution of $(x, i^{\frac{1}{2}}y, i^{\frac{1}{2}}z, t)$ for (x, y, z, t). If at the same time we substitute $(\alpha, i^{\frac{1}{2}}\beta, i^{\frac{1}{2}}\gamma, \delta)$ for $(\alpha, \beta, \gamma, \delta)$ in the equation I a we obtain the equation II a representing a surface having nodes at the points

$$(\alpha, \beta, \gamma, \delta), \quad (i\beta, \alpha, \delta, i\gamma), \quad (i\gamma, \delta, \alpha, i\beta), \quad (\delta, \gamma, \beta, \alpha),$$

and twelve others obtained from these by changing the signs of pairs of coordinates. If α, β, γ, δ are real, eight of these nodes are real.

The substitution of (ix, iy, z, t) for (x, y, z, t) in II a has the effect of interchanging k_5 and k_6, and hence produces II b. It is evident that all the nodes are imaginary. By reasoning similar to that employed in the case of I b it may be deduced that II b is a real surface from the fact that the directrix $x=0$, $y=0$ cuts II a in two real points.

§ 73. TWO REAL AND FOUR IMAGINARY COMPLEXES.

III. In this case two pairs of complexes are imaginary. There is here no subdivision as the question of order of magnitude does not arise. The condition for a real tangent is that $(k_1-\mu_1)(k_1-\mu_2)$ and $(k_2-\mu_1)(k_2-\mu_2)$ must be of the same sign; hence μ_1 and μ_2 may be imaginary, but if real do not occur alternately with k_1 and k_2 and so may be equal. Of thirty-two associated points only eight are real since the signs of conjugate imaginary line coordinates must be changed simultaneously. *There are therefore four real nodes and four real tropes.* Two singular conics cut in two nodes whose parameters are k_1 and k_2, terminating two hyperbolic segments, and the other two conics cut in the other two nodes, so that there are four hyperbolic segments altogether. There are two elliptic segments, each bounded by four arcs of conics and four nodes. The complete surface consists of two pieces each containing two hyperbolic and one elliptic segment, and attached to each other at the four nodes.

Consider the degenerate case when one real node is taken on the quadric 134 . 256. The imaginary directrices (34) belong to one regulus and (56) to the other. The surface is represented in the figure

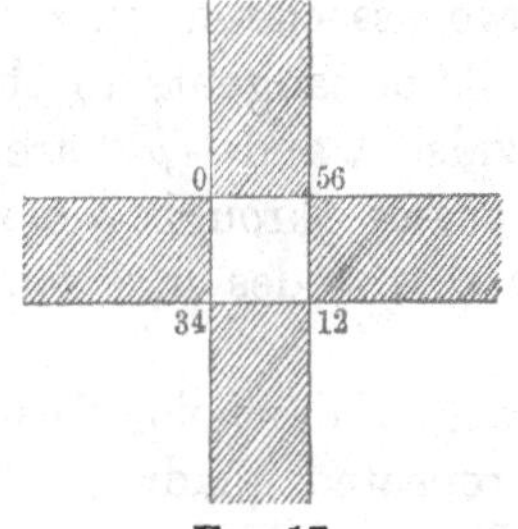

FIG. 17.

showing that there are two pieces attached to each other at four nodes.

The equation of III is obtained from Ia by substituting $(ix, i^{\frac{1}{2}}y, i^{\frac{1}{2}}z, t)$ for (x, y, z, t) and $(i\alpha, i^{\frac{1}{2}}\beta, i^{\frac{1}{2}}\gamma, \delta)$ for $(\alpha, \beta, \gamma, \delta)$. The four real nodes are $(\alpha, \beta, \gamma, \delta)$, $(\alpha, -\beta, -\gamma, \delta)$, $(-\alpha, \beta, -\gamma, \delta)$, $(-\alpha, -\beta, \gamma, \delta)$, forming a Göpel tetrad.

§ 74. SIX IMAGINARY COMPLEXES.

IVa. *Four nodes, two real sheets.*

Lastly we consider the case when the three pairs of complexes are imaginary. There are two subcases. In IVa k_1k_2, k_3k_4, k_5k_6 are pairs of conjugate imaginaries. For a real tangent the line coordinates must be pairs of conjugate imaginaries, and this is the case when the elliptic coordinates $\mu_1\mu_2$ are real or conjugate imaginary. If one (x) of thirty-two associated lines is real then eight are, namely those obtained from $(x_1, x_2, x_3, x_4, x_5, x_6)$ and $(ix_1, -ix_2, ix_3, -ix_4, ix_5, -ix_6)$ by changing the signs of pairs of conjugate imaginary coordinates. Since it is possible to have $\mu_1 = \mu_2$, there are four real nodes and four real tropes, and these are not incident since the parameters of the six nodes on a conic are all imaginary. A hyperbolic segment is funnel-shaped, bounded by one whole conic and one node. Each asymptotic curve consists of four branches one in each hyperbolic segment having a cusp at the node and touching the conic boundary. There are no principal asymptotic curves. There are two elliptic segments, one bounded by the four conics and the other by the four nodes. The wave surface is an example of this case.

IVb. *Four nodes, no real sheets.*

In the second subcase IVb the coefficients k_s have the form

$$r_1e^{i\theta_1}, \quad -r_1^{-1}e^{i\theta_1}, \quad r_3e^{i\theta_3}, \quad -r_3^{-1}e^{i\theta_3}, \quad r_5e^{i\theta_5}, \quad -r_5^{-1}e^{i\theta_5},$$

and if one of thirty-two associated lines is real, then eight are. There are no real pencils of tangents to the surface but the line $(\lambda, \lambda, \mu, \mu)$ is real provided λ and $-\mu^{-1}$ are conjugate imaginaries, and this line either passes through a node or lies in a trope. Hence there are four real nodes and four real tropes, but the surface is imaginary.

As in preceding cases the topology can be inferred from the degenerate case of a repeated quadric.

The equations of IVa and IVb can be obtained by means of the imaginary linear transformation already given. The four real nodes form a Göpel tetrad.

CHAPTER XII.

GEOMETRY OF FOUR DIMENSIONS.

§ 75. LINEAR MANIFOLDS.

By 'geometry of four dimensions' is to be understood a method of reasoning about sets of numbers and equations, in which the principles of elementary algebra are clothed in a language analogous to that of ordinary geometry. Although we cannot bring our intuition to bear directly upon four-dimensional configurations we can do so indirectly by creating an artificial intuition based on analogy. Many theorems in algebra are readily extended from three variables to four and the new theorems are expressed by an extension of ordinary geometrical nomenclature. Thus when we have learned the laws of extension we can reason in the new nomenclature without being able to attach an actual geometrical significance to the terms used.

A set of five coordinates x_1, x_2, x_3, x_4, x_5, used homogeneously, is called a *point* (x). It may be determined by four independent linear equations.

Two points (x) and (y) determine a single infinitude of points $(\lambda x + \mu y)$ called a *line*. Three points (x) (y) and (z) determine the ∞^2 points $(\lambda x + \mu y + \nu z)$ of a *plane* provided (z) is not determined by (x) and (y), that is, if the three points are not collinear. Similarly four points determine the ∞^3 points of a *space* provided they are linearly independent, that is, if they do not lie in a plane. On eliminating λ, μ, ν, ρ from the five equations

$$w_s = \lambda x_s + \mu y_s + \nu z_s + \rho t_s \qquad (s = 1, 2, 3, 4, 5)$$

we obtain a single relation linear in the coordinates w_s, so that a space consists of all the points whose coordinates satisfy one linear equation. The line, plane, and space are linear *manifolds* of one, two, and three dimensions respectively.

The following table gives the number of equations and points required for the determination of the various elements:

element	equations	points
point	4	1
line	3	2
plane	2	3
space	1	4

As in three dimensions, so here an *intersection* means a common point or set of points, that is a common solution or set of solutions of certain equations. By adding the numbers in the middle column of the preceding table we can construct the following table of intersections, all the elements being supposed to have general positions.

	line	plane	space
line			point
plane		point	line
space	point	line	plane

Lastly, it is important to realise the manifold of smallest dimensions which contains two given manifolds in general position: this is shown in the following table which is constructed by adding the numbers in the last column of the first table.

	point	line	plane
point	line	plane	space
line	plane	space	
plane	space		

The meaning of *projection*, like that of intersection, may be extended to space of higher dimensions. If A, B, C are manifolds, to project B from A on to C means to construct the smallest manifold containing both A and B and then find its intersection with C. Thus if A is a point and C a space the projection of B has the same dimensions as B; so too if A is a line and C a plane*.

§ 76. CONSTRUCTION OF THE 15_6 CONFIGURATION FROM SIX POINTS IN FOUR DIMENSIONS†.

We now proceed to show that the general 16_6 configuration in ordinary space can be obtained by the operations of section and projection from the figure of six points in space of four dimensions.

Let the points be called 1, 2, 3, 4, 5, 6. They determine fifteen lines which may be called 12, ..., and twenty planes 123 ..., and fifteen spaces 1234.... The line 12 cuts the *opposite* space 3456 in a *diagonal point* P_{12}. These fifteen diagonal points lie by threes on fifteen lines, for P_{12}, P_{34}, P_{56} lie in each of the spaces 3456, 1256, 1234, and any three spaces have a line in common. These are called *transversal lines* and may be denoted by (12.34.56), etc. Thus the line (12.34.56) contains the points P_{12}, P_{34}, P_{56}. Again the three lines (12.34.56), (12.35.46), (12.45.36) meet in the point P_{12} and are contained in the same space 3456.

Corresponding to a partition of the six points into two sets of three, e.g. 123.456, we get nine diagonal points

$$\begin{matrix} P_{14} & P_{25} & P_{36} \\ P_{26} & P_{34} & P_{15} \\ P_{35} & P_{16} & P_{24} \end{matrix}$$

which are seen to lie upon six transversal lines corresponding to the rows and columns of the scheme. Hence the space determined by any four of these points which are linearly independent contains the remaining points and is called a *cardinal space* and denoted by (123.456). It follows that the six transversals are three generators of one system and three generators of the other system of a quadric surface lying in

* Veronese-Schepp, *Grundzüge der Geometrie von mehreren Dimensionen.*

† Richmond, *Quarterly Journal*, XXXI, 125; XXXIV, 117.

the cardinal space. There are ten cardinal spaces and the fifteen transversal lines lie by sixes on ten quadrics contained in them.

Consider now the three-dimensional figure obtained by cutting the transversal lines by an arbitrary space. A cardinal space is cut in a plane and the six transversal lines in it in six points lying on a conic. We have therefore a 15_6 configuration of fifteen points lying by sixes on ten conics. This is the configuration of nodes and tropes of a general fifteen-nodal quartic surface, and we shall see presently that this surface is the section by an arbitrary space of a certain quartic *variety* or curved threefold in space of four dimensions. Further, when the section is by a *tangent* space, one more node appears and we have the general Kummer surface.

§ 77. ANALYTICAL METHODS.

In space of four dimensions a point is represented by five homogeneous coordinates, but it is convenient to use six, x_1, x_2, x_3, x_4, x_5, x_6 connected by the relation

$$x_1 + x_2 + x_3 + x_4 + x_5 + x_6 = 0.$$

The equation of any space is

$$u_1x_1 + u_2x_2 + u_3x_3 + u_4x_4 + u_5x_5 + u_6x_6 = 0$$

which is unaltered when the same quantity is added to the coefficients. To fix their values we assume

$$u_1 + u_2 + u_3 + u_4 + u_5 + u_6 = 0.$$

The coordinates may be so chosen that the equations of the point 1 are $u_1 = 0$ in space coordinates and $x_2 = x_3 = x_4 = x_5 = x_6$ in point coordinates, and similarly for the other points.

Hence the space 3456 has coordinates $(1, -1, 0, 0, 0, 0)$ and equation $x_1 = x_2$; from this it may easily be proved that the diagonal point P_{12} is

$$u_1 + u_2 = 0; \quad x_1 = x_2, \quad x_3 = x_4 = x_5 = x_6;$$

the transversal line (12.34.56) is

$$u_1 + u_2 = u_3 + u_4 = u_5 + u_6 = 0; \quad x_1 = x_2, \ x_3 = x_4, \ x_5 = x_6$$

and the cardinal space (123.456) is

$$u_1 = u_2 = u_3 = -u_4 = -u_5 = -u_6; \ x_1 + x_2 + x_3 = x_4 + x_5 + x_6 = 0.$$

An arbitrary space

$$a_1x_1 + a_2x_2 + a_3x_3 + a_4x_4 + a_5x_5 + a_6x_6 = 0$$

cuts the transversal line (12.34.56) where

$$(a_1 + a_2)\, x_1 + (a_3 + a_4)\, x_3 + (a_5 + a_6)\, x_5 = 0$$

so that the equation of the point of intersection can be expressed in the equivalent forms

$$\frac{u_1 + u_2}{a_1 + a_2} = \frac{u_3 + u_4}{a_3 + a_4} = \frac{u_5 + u_6}{a_5 + a_6}.$$

In this way we have a symmetrical expression of the 15_6 configuration.

§ 78. THE 16_6 CONFIGURATION.

It is possible in six ways to select a group of five transversal lines which contain all the diagonal points and of which no three belong to the same cardinal space; in fact these groups correspond to the six pairs of cyclical arrangements of five figures, or to the six pairs of mutually inscribed pentagons whose edges are the intersections of five planes (p. 10). Thus, corresponding to 23456 or 24635 we have the group of transversal lines

12.36.45

13.42.56

14.53.62

15.64.23

16.25.34.

We proceed to prove that when $\Sigma a_s^3 = 0$ the five points of intersection of these lines with the space $\Sigma a_s x_s = 0$ are coplanar. The plane containing the first three points is determined by the three tangential equations

$$\frac{u_1 + u_2}{a_1 + a_2} = \frac{u_3 + u_6}{a_3 + a_6}, \quad \frac{u_2 + u_4}{a_2 + a_4} = \frac{u_6 + u_5}{a_6 + a_5}, \quad \frac{u_4 + u_1}{a_4 + a_1} = \frac{u_5 + u_3}{a_5 + a_3}.$$

Now in consequence of $\Sigma a_s = 0$, $\Sigma a_s^3 = 0$, we have

$$(a_1 + a_2)\,(a_2 + a_4)\,(a_4 + a_1) + (a_3 + a_6)\,(a_6 + a_5)\,(a_5 + a_3) = 0$$

whence, by the preceding equations,

$$(u_1 + u_2)\,(u_2 + u_4)\,(u_4 + u_1) + (u_3 + u_6)\,(u_6 + u_5)\,(u_5 + u_3) = 0$$

leading to $\Sigma u_s^3 = 0$, and the symmetry of this result shows that the plane through the first three points passes through the other two. Hence when the single condition $\Sigma a_s^3 = 0$ is satisfied, the configuration of fifteen diagonal points contains six sets of five coplanar points; in addition to the ten sets of six points each on a conic.

One of the last six planes is the intersection of the two spaces $\Sigma a_s x_s = 0$, $\Sigma u_s x_s = 0$, the u_s being given by the preceding equations. It is evident that those equations are still satisfied if the u_s are replaced by $u_s + \lambda a_s$, λ being arbitrary. It follows that $\Sigma u_s a_s^2 = 0$, showing that the six new planes meet in the point of which this is the tangential or space equation, so that this point completes, with the fifteen, the Kummer configuration of sixteen points and sixteen planes*.

§ 79. GENERAL THEORY OF VARIETIES.

The name *variety* is here given to the threefold locus of ∞^3 points which satisfy a single relation. The simplest variety is when the relation is linear and then the special name *space* is used, as an abbreviation for *linear threefold space*.

Let the equation of a variety in four non-homogeneous coordinates be

$$f(x, y, z, t) = 0,$$

then in the neighbourhood of any point (x', y', z', t') of it, a first approximation is given by

$$(x - x')\,\partial f/\partial x' + (y - y')\,\partial f/\partial y' + (z - z')\,\partial f/\partial z' + (t - t')\,\partial f/\partial t' = 0,$$

which, by analogy, is called the *tangent space* at the point (x', y', z', t').

Change the coordinates so that the origin is the point considered, and $t = 0$ is the tangent space; then, on expanding f in a series of homogeneous polynomials, the equation of the variety is

$$0 = t + f_2(x, y, z, t) + f_3 + \ldots$$

and it is seen at once that the section by the tangent space $t = 0$ is an ordinary surface in that space having a node at the origin.

If x', y', z' are *any* small quantities of the first order, $(x', y', z', 0)$ is, to this order, a near point on the variety and the tangent space is

$$x\partial f_2/\partial x' + y\partial f_2/\partial y' + z\partial f_2/\partial z' + t = 0,$$

small quantities of the second order being omitted. This cuts the space $t = 0$ in the polar plane of the point (x', y', z') with respect

* Similar analytical treatment can be applied to the theories of lines on a cubic surface, bitangents of a plane quartic, and Pascal's figure. See Richmond, *Camb. Phil. Trans.* xv, 267. Cremona, *Math. Ann.* XIII, 301.

to the cone $f_2(x, y, z, 0) = 0$. Thus all the ∞^2 planes which lie in a tangent space and pass through the point of contact may be called *tangent planes* and cut the variety in a curve having a double point. A finite number of tangent spaces, equal to the *class* of the variety, pass through an arbitrarily given plane, and in the case of a tangent plane, two of the tangent spaces coincide.

The points in which any line $x/x' = y/y' = z/z' = t/t'$ cuts the variety are given by the equation in k

$$f(kx', ky', kz', kt') = 0.$$

The degree of this equation is the *order* of the variety and is equal to the order of any space section. One root is zero, and a second is zero if $t' = 0$. Hence all the ∞^2 lines in a tangent space which pass through the point of contact may be called *tangent lines*. Of these, ∞^1 have three-point contact and are the generators of a quadric cone $f_2 = 0$, $t = 0$, and of these six have four-point contact, being the intersections of the two cones

$$t = 0, \quad f_2(x, y, z, 0) = 0, \quad f_3(x, y, z, 0) = 0.$$

All these theorems can be reciprocated. If the general equation of a space is

$$lx + my + nz + p + qt = 0$$

a single equation

$$\phi(l, m, n, p, q) = 0$$

may be regarded as the tangential equation of a variety in space coordinates $l : m : n : p : q$. It is unnecessary to repeat all the theorems. The equation of the point of contact of the tangent space (l', m', n', p', q') is

$$l\,\partial\phi/\partial l' + m\,\partial\phi/\partial m' + n\,\partial\phi/\partial n' + p\,\partial\phi/\partial p' + q\,\partial\phi/\partial q' = 0.$$

A *singular point* of a variety is one at which the tangent space is indefinite. The conditions that (x, y, z, t) may be a singular point of $f = 0$ are

$$\partial f/\partial x = 0, \quad \partial f/\partial y = 0, \quad \partial f/\partial z = 0, \quad \partial f/\partial t = 0.$$

Thus if the origin is a singular point the equation has the form

$$0 = f_2(x, y, z, t) + f_3 + \dots.$$

The section by every space through the origin is a surface having a node. ∞^2 lines through the origin have three-point contact.

A locus of ∞^1 singular points is a *singular line*. Let the tangent to this line be $x=y=z=0$. Then we must have $\partial f_2/\partial t=0$ and the variety is

$$0=f_2(x, y, z)+f_3(x, y, z, t)+\ldots.$$

The section by $lx+my+nz=0$ is a surface for which the nodal cone at the origin breaks into two planes.

A *singular tangent space* is one whose point of contact is indefinite. Taking this to be $t=0$, and the general equation of a plane to be

$$lx+my+nz+t+p=0$$

the tangential equation of the variety must be

$$0=\phi_2(l, m, n, p)+\phi_3+\ldots.$$

The first approximation, instead of being linear and giving a single point of contact, is now quadratic,

$$\phi_2(l, m, n, p)=0,$$

and represents a quadric surface. We may say that the singular tangent space has contact at all the points of a quadric surface. We therefore infer that this surface appears repeated in the complete section by the tangent space.

§ 80. SPACE SECTIONS OF A CERTAIN QUARTIC VARIETY.

Returning to the six homogeneous space coordinates u_s connected by the relation

$$u_1+u_2+u_3+u_4+u_5+u_6=0$$

consider the variety of the third class whose tangential equation is

$$u_1^3+u_2^3+u_3^3+u_4^3+u_5^3+u_6^3=0.$$

In consequence of $\Sigma u_s=0$ this can be written in the form

$$(u_2+u_3)(u_3+u_1)(u_1+u_2)+(u_5+u_6)(u_6+u_4)(u_4+u_5)=0$$

and in nine other similar forms. This equation is satisfied by

$$u_1+u_2=0, \quad u_5+u_6=0,$$

that is to say, by every space containing the transversal line (12.34.56). Hence the ∞^2 spaces through this line are all tangent spaces, and of them ∞^1 have a given point of it for point of contact. On taking an arbitrary space section we get a point on a surface at which there are ∞^1 tangent planes; hence the point is a node on the section. Thus we see that the section of the variety by an arbitrary space is a surface having fifteen nodes, the sections of the transversal lines; these have been shown to lie by sixes on ten conics, the sections of the cardinal spaces.

It follows immediately from the general theory that the point of contact of a tangent space $\Sigma(u_s x_s) = 0$ has coordinates whose differences are proportional to $u_r^2 - u_s^2$, etc. Hence the only singular tangent spaces are given by

$$u_1^2 = u_2^2 = u_3^2 = u_4^2 = u_5^2 = u_6^2$$

which have ten possible solutions of which one is

$$u_1 = u_2 = u_3 = -u_4 = -u_5 = -u_6$$

or the cardinal space (123.456). Hence each cardinal space touches the variety at all the points of a quadric surface, and in the arbitrary space section we get a plane touching the fifteen-nodal surface along a conic containing six nodes. We recognise that the section must be a quartic surface with fifteen nodes and ten tropes.

By taking a tangent space for the space of section we get a new node at the point of contact. The section is now a Kummer surface and therefore six new tropes must appear.

Now in a tangent space (u) to a variety there are six planes through the point of contact such that four consecutive spaces through them are tangent spaces. In the present case the variety is of class three and so *every* space $(u + \lambda v)$ through any one of these six planes is a tangent space. The conditions are $\Sigma u^3 = 0$, $\Sigma u^2 v = 0$, $\Sigma uv^2 = 0$, $\Sigma v^3 = 0$ and show that the point of contact of $(u + \lambda v)$ lies in (u) and therefore in the plane. This plane has therefore ∞^1 points of contact with the section, and in this way the additional tropes are accounted for.

The point equation of the variety $\Sigma u_s^3 = 0$ can be written in a great many ways. One is

$$\{(x_1 - x_2)^2 - (x_3 + x_4 - x_5 - x_6)^2\}^{\frac{1}{2}} + \{(x_3 - x_4)^2 - (x_5 + x_6 - x_1 - x_2)^2\}^{\frac{1}{2}} + \{(x_5 - x_6)^2 - (x_1 + x_2 - x_3 - x_4)^2\}^{\frac{1}{2}} = 0$$

in which no use is made of the linear relation which connects the x_s. There are fifteen equations of this type. It may be written

$$(xx')^{\frac{1}{2}} + (yy')^{\frac{1}{2}} + (zz')^{\frac{1}{2}} = 0$$

where $x = 0 \ldots z' = 0$ are the equations of six cardinal spaces and the identical relation connecting them is

$$x + y + z + x' + y' + z' = 0 \quad \ldots\ldots\ldots\ldots\ldots\ldots (1).$$

From the point equation of the greater variety it is evident that the cardinal spaces are singular tangent spaces. The general tangent space is

$$(x_0'/x_0)^{\frac{1}{2}} x_0 + (x_0/x_0')^{\frac{1}{2}} x_0' + (y_0'/y_0)^{\frac{1}{2}} y_0 + (y_0/y_0')^{\frac{1}{2}} y_0' + (z_0'/z_0)^{\frac{1}{2}} z_0 + (z_0/z_0')^{\frac{1}{2}} z_0' = 0 \text{............(2)},$$

where $$(x_0x_0')^{\frac{1}{2}} + (y_0y_0')^{\frac{1}{2}} + (z_0z_0')^{\frac{1}{2}} = 0.$$

Equations (1) and (2) are of the same kind as those of § 55 which must connect the six tropes of a fourteen-nodal quartic surface in order that two additional nodes may appear. The present chapter gives an interpretation of those equations by means of space of higher dimensions.

By starting with the reciprocal variety and proceeding by reciprocal processes we shall arrive at reciprocal results. On account of the duality of Kummer's surface we reach it again by this method. Thus instead of taking a space section of the quartic variety, we may take the reciprocal cubic variety and the enveloping 'cone' from any point. Thus any space section of this 'cone' is a surface having ten nodes and fifteen tropes, the reciprocal of the fifteen-nodal quartic surface. When the vertex of the 'cone' is on the variety, the section is a Kummer surface.

CHAPTER XIII.

ALGEBRAIC CURVES ON THE SURFACE.

§ 81. GEOMETRY ON A SURFACE.

In preceding chapters Kummer's surface has been considered as a figure in space of three dimensions, mainly in relation to various systems of points, lines, and planes. The surface has been treated as a whole, being uniquely determined by its singularities, and these form a configuration which is conveniently studied first and independently. Now, however, we must turn our attention to the surface as a two-dimensional field of geometry, and consider the curves which can be traced upon it. Here a further subdivision of the subject arises according as we investigate the curves in the neighbourhood of a particular point, or treat them in their entirety, the former branch is especially devoted to transcendental curves and those defined by differential equations: the latter to algebraic curves.

An important step is made in the theory of an algebraic surface when the coordinates are expressed as *uniform* functions of two parameters, for we are then able to transfer theorems in plane geometry to the surface. Every curve on the surface has an equation, expressing a relation between these parameters. The properties of a surface depend largely upon the kind of function which must be employed in the parametric expression, and a detailed study of these functions is therefore necessary.

But there is another method of investigating *algebraic* curves based upon their characteristic property of cutting every algebraic surface in a finite number of points; for a surface of sufficiently high order can always be found to contain the whole of such a curve, and the curve may therefore be defined as the part common to several surfaces. In this way a curve is defined by several equations taken together, which have the advantage of being algebraic. The chief difficulty in the theory is that in general two equations do not suffice, for a given curve may not be the *complete* intersection of any two surfaces; the simplest example of this is the twisted cubic. It is possible, however, to arrange that two surfaces may pass through the curve and cut again in straight lines only, as in Cayley's representation by means of a cone and a *monoid**. The given curve is projected from the point $x=y=z=0$ by a cone $f(x, y, z)=0$, the vertex being chosen so that not an infinite number of chords pass through it. Then arbitrary values of $x:y:z$ determine *one* value of the remaining coordinate, which must therefore be given by an equation of the form

$$t\chi(x, y, z)=\psi(x, y, z)$$

representing a monoid surface. The complete intersection consists of the given curve and any lines which may be common to the cones $f=0$, $\chi=0$, $\psi=0$.

The theory of algebraic curves on the *general* Kummer surface is simplified by the remarkable theorem that *a surface can be found to touch Kummer's surface all along any given algebraic curve lying thereon and have no further intersection with the surface*; the curve, counted twice, is the *complete* intersection of two surfaces and can therefore be represented by the equation of the tangent surface alone. When the Kummer surface is not perfectly general it may happen that curves exist on it for which the theorem is not true.

§ 82. ALGEBRAIC CURVES ON KUMMER'S SURFACE.

Let the equation of Kummer's surface Φ be written as in § 10 in the form

$$\phi_2 t^2+2\phi_3 t+\phi_4=0,$$

so that the point $x=y=z=0$ is a node at which the tangent cone is $\phi_2=0$. Any curve on the surface can be represented as the intersection of a cone

$$f(x, y, z)=0$$

and a monoid
$$t\chi(x, y, z)=\psi(x, y, z)$$

after excluding the lines common to $f=0$, $\chi=0$, $\psi=0$. If f were a general polynomial in its arguments the curve of intersection with Φ would have the special property of cutting each generator

* *Collected Papers*, v, 7.

of the cone *twice*, and the monoid representation would fail. It is required to find what special form f must have in order that the complete intersection may break up into two, projections of each other from the node.

On solving the quadratic for t we find

$$\phi_2 t + \phi_3 = \pm\sqrt{\phi_3^2 - \phi_2\phi_4}$$

so that, in the language of two dimensions, $f=0$ is a curve upon which $\sqrt{\phi_3^2 - \phi_2\phi_4}$ has a rational value in terms of the point coordinates, namely $\phi_2\psi/\chi + \phi_3$; and the problem of finding all the algebraic curves on the surface is the same as of finding all plane curves $f=0$ having the preceding property*. A necessary and sufficient condition is that f must be a *factor* of an expression of the form $F^2 - G^2(\phi_3^2 - \phi_2\phi_4)$, where F and G are homogeneous polynomials in x, y, z; but this does not define the form of f with sufficient precision. Now $\phi_3^2 - \phi_2\phi_4$ is the product of six linear factors $x_1 x_2 x_3 x_4 x_5 x_6$; let X be the product of some of these factors and Y the product of the rest, then f must be a factor of an expression of the form $F^2X - H^2Y$.

Since the sextic $\phi_3^2 - \phi_2\phi_4 = 0$ is a hexagram circumscribing the conic $\phi_2 \equiv y^2 - xz = 0$, we adopt a new system of coordinates having special reference to this conic and substitute $1, \frac{1}{2}(u+v), uv$, for x, y, z respectively. Then (p. 18),

$$x_s = (u - k_s)(v - k_s)$$

and $$\phi_3^2 - \phi_2\phi_4 = \Pi x_s = \Pi(u-k_s)(v-k_s) = \Pi \text{ say.}$$

Let U be the product of six of the twelve factors of Π involving all the six k_s and let V be the product of the remaining factors, so that U becomes V when u and v are interchanged; further let P be any polynomial in u and v and let Q be what P becomes when u and v are interchanged. Then the equation

$$\frac{P^2U - Q^2V}{u - v} = 0$$

is integral and symmetric in u, v, and therefore represents a plane algebraic curve upon which $\sqrt{UV}$ has the value $(P^2U + Q^2V)/2PQ$, and this value, being symmetric in u, v, is rational in x, y, z.

It is a remarkable theorem, and one not easy to prove directly, that every curve upon which $\sqrt{x_1 x_2 x_3 x_4 x_5 x_6}$ is rational is expressible *exactly* in one of these two forms, that is, without

* Hudson, *Math. Gazette*, July, 1904, p. 56.

extraneous factors, *provided* $k_1, k_2, k_3, k_4, k_5, k_6$ *are perfectly general.* Expressing this differently, none of the curves

$$F^2X - G^2Y = 0,$$

or

$$\frac{P^2U - Q^2V}{u - v} = 0$$

is reducible except into curves whose equations have one or other of these two forms. At present we assume this theorem in order to be able to enunciate subsequent propositions with complete generality. It can however be verified for the curves of the different orders taken in turn.

The equation $P^2U - Q^2V = 0$ represents two curves on the surface which are distinguished by the pairs of equations

$$\left.\begin{array}{r} P\sqrt{U} + Q\sqrt{V} = 0 \\ \phi_2 t + \phi_3 = \sqrt{UV} \end{array}\right\} \text{ and } \left.\begin{array}{r} P\sqrt{U} + Q\sqrt{V} = 0 \\ \phi_2 t + \phi_3 = -\sqrt{UV} \end{array}\right\}.$$

In what follows the equation of the surface will be understood to be given in a *definite* irrational form, and then a single equation $P\sqrt{U} + Q\sqrt{V} = 0$ determines a single curve on the surface.

We have next to show that the equation of any curve can be expressed in terms of products (with repetitions) of the sixteen radicals $\sqrt{x_s}$, $\sqrt{x_{rst}}$, where, as on p. 19,

$$x_s = (u - k_s)(v - k_s) = u_s v_s \qquad (s = 1, 2, 3, 4, 5, 6),$$

$$x_{123} = x_{456} = (\sqrt{u_1 u_2 u_3 v_4 v_5 v_6} + \sqrt{v_1 v_2 v_3 u_4 u_5 u_6})^2/(u - v)^2.$$

This is evident for the symmetrical equation $F\sqrt{X} + G\sqrt{Y} = 0$. The expression is not unique on account of numerous identities among the radicals, of which an example is

$$(u - v)^2 \sqrt{x_{123} x_{124}} = (u_1 u_2 v_5 v_6 + v_1 v_2 u_5 u_6)\sqrt{u_3 u_4} + (u_3 v_4 + v_3 u_4)\sqrt{x_1 x_2 x_5 x_6}.$$

The case of the equation $P\sqrt{U} + Q\sqrt{V} = 0$ may be illustrated by an example. Let

$$U = u_1 v_2 v_3 v_4 v_5 v_6, \quad V = v_1 u_2 u_3 u_4 u_5 u_6,$$

then

$$(u - v) \sum_{234} (k_2 - k_3) \sqrt{x_2 x_3 x_{123}} = -(k_2 - k_3)(k_3 - k_4)(k_4 - k_2)(\sqrt{U} + \sqrt{V}),$$

$$(u - v) \sum_{234} (k_2 - k_3) k_4 \sqrt{x_2 x_3 x_{123}}$$
$$= -(k_2 - k_3)(k_3 - k_4)(k_4 - k_2)(u\sqrt{U} + v\sqrt{V}),$$

and so on, whence the general pair of terms in $P\sqrt{U} + Q\sqrt{V}$ can be found.

§ 83. THE Θ-EQUATION OF A CURVE.

With the assumption made at the end of § 82 it has been shown that, when the equation of the general Kummer surface Φ is given in an irrational form with a definite sign to the radical, *every* algebraic curve on Φ is expressible by an equation of the form

$$\lambda_1 \sqrt{p_1} + \lambda_2 \sqrt{p_2} + \ldots = 0,$$

where p_s is a product of some of the sixteen linear forms here denoted by x_s, x_{rst}, and λ_s is numerical*. We shall generally use Θ to denote the left side of this equation, so that $\Theta \equiv \Sigma \lambda_s p_s$. Θ is distinguished by the following three properties.

From the nature of the coordinates used, Θ must be homogeneous in the linear forms. The number of factors in any product p_s is called the *order* of Θ, or of the equation.

After substitution for the radicals in terms of u and v every term of the equation takes the form $P\sqrt{U} + Q\sqrt{V}$, U and V being the same for all the terms. If the equation is of the first kind, formerly written $F\sqrt{X} + G\sqrt{Y} = 0$, U and V are symmetric in u and v and hence p_s contains an *even* number of factors of the type x_{rst}; if the equation is of the second kind U becomes V when u and v are interchanged, and p_s contains an *odd* number of factors of the type x_{rst}. Accordingly Θ is distinguished by its *parity*, being even or odd. The parity of a product and of an equation is thus defined in reference to a particular set of six elements $x_1, x_2, x_3, x_4, x_5, x_6$. As a rule it is not the same for all the sixteen sets, that is, it is not an invariant property under the group of sixteen collineations, and accordingly does not imply any essential geometrical distinction. It is, however, an important feature when it is invariant, as for instance when the factors of a product form a Göpel tetrad such as $x_1 x_2 x_{156} x_{256}$ which is even, or a Rosenhain tetrad such as $x_1 x_2 x_3 x_{123}$ which is odd (see pp. 78, 79). It is customary in estimating parity to count the number of factors of the type x_s instead of x_{rst}, leading to a different result when the order is odd.

Thirdly, we have to consider the property that after substitution in terms of u and v every term of Θ involves the same pair of radicals $\sqrt{U}$ and $\sqrt{V}$. Now one of the radicals in the case of

* It would involve no assumption and might therefore be more satisfactory to start with equations of this kind and show that they represent algebraic curves. But we could not be sure of dealing with every curve in this way. To complete the theory transcendental methods are required (cf. § 104, below).

$\sqrt{x_{123}}$ is the same as in $\sqrt{x_1 x_2 x_3}\sqrt{u_1 u_2 u_3 u_4 u_5 u_6}$ from which it follows, both for even and for odd equations, that if in the terms of Θ every factor of type x_{rst} be replaced by the corresponding product $x_r x_s x_t$, all the terms will contain one of two irrationalities whose product is $\sqrt{x_1 x_2 x_3 x_4 x_5 x_6}$. This gives an important rule for finding which products may be associated in the same equation; it may be expressed in another way. The suffixes of the sixteen linear forms are the symbols of dualistic transformations obtaining the tropes from one node; the laws of compounding these transformations are simply that every two are permutable and that

$$11 = 22 = 33 = 44 = 55 = 66 = 123456 = \text{identity}.$$

Accordingly we have the rule that if the operations represented by the suffixes of the factors in any term of Θ are compounded, the resulting operation is the same for every term. The symbol of this operation is called the *characteristic* of the product, and of Θ, and of the equation $\Theta = 0$. Of the sixteen characteristics there is one which is distinguished from the rest, namely the symbol of identity, *dd*, or 0. The remaining fifteen characteristics may be treated alike and will be denoted by two-letter or two-figure symbols. It will be necessary to speak of the parity of a characteristic $\alpha\beta$, and by this is meant the parity of $x_{\alpha\beta}$. An equation of order n and characteristic $\alpha\beta$ is written

$$\overset{(n)}{\underset{\alpha\beta}{\Theta}} = 0.$$

When the order, parity, and characteristic are given the equation is said to belong to a certain *family*. In order to construct the general equation of a given family we pick out from among the sixteen radicals all possible products having the given order, parity, and characteristic, retain only those which are linearly independent on the surface, and combine these linearly with undetermined coefficients.

§ 84. GENERAL THEOREMS ON CURVES.

Let $p_1, p_2 \dots$ be all possible products of n of the sixteen linear forms satisfying the conditions of having given parity and characteristic; we consider the family of curves represented by the equation of order n

$$\Theta \equiv \lambda_1 \sqrt{p_1} + \lambda_2 \sqrt{p_2} + \dots = 0.$$

Let $\Theta = 0$ and $\Theta' = 0$ be the equations of two curves of the family. The product $\Theta\Theta'$ is rational except for terms of the form $(\lambda_1\lambda_2' + \lambda_2\lambda_1')\sqrt{p_1p_2}$; but if we replace the coordinates by their expressions in terms of u and v the irrational part of $\sqrt{p_1p_2}$ is the same as $\sqrt{x_1x_2x_3x_4x_5x_6}$ which is rational and integral on the surface Φ. Hence the equation $\Theta\Theta' = 0$ can be rationalised by means of the equation $\Phi = 0$, and then represents an algebraic surface of order n cutting Φ in the curves $\Theta = 0$, $\Theta' = 0$ only. Hence we have the theorem:

Any two curves of the same family form the complete intersection with a surface of order n.

By making the two curves coincide we infer:

Every curve is the curve of contact and sole intersection with a tangent surface of order n.

The equation of the tangent surface is $\Theta^2 = 0$, after rationalising by means of $\Phi = 0$. This is of order n and therefore the complete intersection is of order $4n$; hence

Every curve is of even order $2n$.

A $2m$-ic curve meets a $2n$-ic curve where it meets the n-ic surface tangent along the latter. The $2mn$ intersections of curve and surface all lie on Φ and are contacts except at nodes. Hence *any two curves of orders $2m$ and $2n$ intersect at an even number, $2p$, of nodes and at $mn - p$ other points.*

Let $S_1 = 0$, $S_2 = 0$ be the surfaces of order n, tangent along two curves $\Theta_1 = 0$, $\Theta_2 = 0$ of the same family, and let $S = 0$ be the surface of the same order containing both curves; then, in virtue of $\Phi = 0$,

$$\Theta_1^2 = S_1, \quad \Theta_2^2 = S_2, \quad \Theta_1\Theta_2 = S,$$

whence $S_1S_2 = S^2$ in virtue of $\Phi = 0$, and therefore we have the identity

$$S_1S_2 \equiv S^2 + G\Phi,$$

$G = 0$ being a surface of order $2n - 4$, touching S_1 and S_2 along the remaining intersections with S. From this identity many remarkable results can be deduced.

The first two theorems of this section are particular cases of the following:

Every equation of even order, even, and of zero characteristic is rational on the surface and represents a complete intersection.

For, from the first two qualifications it follows that the number of factors of type $\sqrt{x_{123}}$ in any term is even; hence the irrationality is the same as when this factor is replaced by $\sqrt{x_1x_2x_3}$ or by $\sqrt{x_4x_5x_6}$. But every product of $\sqrt{x_1} \ldots \sqrt{x_6}$ which has zero characteristic is either rational or is $\sqrt{x_1x_2x_3x_4x_5x_6}$ which is rational on Φ.

From this theorem is deduced another which greatly facilitates the study of curves on the surface, namely:

Every curve and not more than four singular conics together form a complete intersection.

For, if $\Theta = 0$ is the equation of any curve, it can be rationalised by finding a product p of the same parity and characteristic whose order has the same parity as that of Θ; then $\Theta\sqrt{p} = 0$ is an equation of even order, even, and of zero characteristic and therefore, by the preceding theorem, is rational on Φ and represents an algebraic surface cutting Φ in the curve $\Theta = 0$ and the singular conics in the tropes $p = 0$. We shall show in the next section, by examining all the different cases that arise, that p need not contain more than four factors.

Thus when a family of curves is given by the equation

$$\Theta \equiv \Sigma\, \lambda_s \sqrt{p_s} = 0$$

the first step in the investigation is to find a product p of least order such that, in virtue of $\Phi = 0$, $\sqrt{pp_s} = P_s$, a rational integral function of the coordinates: then the family of curves is cut out by the family of surfaces

$$\lambda_1 P_1 + \lambda_2 P_2 + \ldots = 0$$

having for base curves the singular conics in the tropes $p = 0$. The curves therefore form a *linear system* whose *dimension* is one less than the number of linearly independent P_s, after making use of $\Phi = 0$.

The surface $\Sigma\lambda_s P_s = 0$ passes through all the nodes in the planes $p = 0$, and, for general values of λ_s, through no others. Since the tangent cone at a node is of the second order, the complete intersection passes an even number of times through a node and hence the curve $\Theta = 0$ passes through only the nodes common to an odd number of the tropes $p = 0$. Thus all the curves of a family pass through the same nodes, which are the only base points of the linear system.

§ 85. CLASSIFICATION OF FAMILIES OF CURVES.

We shall now examine all the different kinds of equations and the least products which are required to rationalise them, and shall show that the number of factors in this product does not exceed four.

When the order is given, there are thirty-two different families, for with each of the sixteen characteristics the equation may be even or odd. We shall find that families of even order are of three distinct kinds and those of odd order are of only two distinct kinds.

First let the order n be even and the characteristic zero. If $\Theta^{(n)}$ is even it has been proved to be rational on Φ, and the curves are the complete intersections with surfaces of order $\frac{1}{2}n$, and have no base points, and in general pass through no nodes.

If Θ is odd the factors of p form an odd or Rosenhain tetrad, for example $x_1x_2x_3x_{123}$, and then $\Theta\sqrt{p}=0$ represents the complete intersection with a surface of order $\frac{1}{2}n+2$ passing through four conics. Since an odd number of the planes $p=0$ pass through each node, the curves of the family pass through all the nodes.

If the characteristic is not zero it determines two associated octads: each consists of four pairs, giving four even and four odd products of order two and the same characteristic. $\Theta_{\alpha\beta}$ may be rationalised by means of any one of the four products of the same parity, and then we have surfaces of order $\frac{1}{2}n+1$ passing through two conics. These two conics have two common nodes and the remaining nodes on them form an octad and are the base points of the system.

Secondly let the order be odd. In this case all sixteen characteristics may be treated alike but a difference arises according as $\Theta_{\alpha\beta}$ and $x_{\alpha\beta}$ have the same parity or not. If they are both even or both odd, $\Theta_{\alpha\beta}\sqrt{x_{\alpha\beta}}=0$ is rational on Φ and represents the complete intersection with a family of surfaces of order $\frac{1}{2}(n+1)$ passing through one conic. The six nodes on this conic are the base points of the system of curves.

If $\Theta_{\alpha\beta}$ and $x_{\alpha\beta}$ have opposite parity, the factors of the product p are three linear forms which with $x_{\alpha\beta}$ make up a Rosenhain tetrad. Then since $px_{\alpha\beta}$ is odd and has zero characteristic it follows that p and $x_{\alpha\beta}$ have opposite parity, and $\Theta_{\alpha\beta}\sqrt{p}=0$ is rational, and represents the complete intersection with a family of surfaces of order $\frac{1}{2}(n+3)$ passing through three conics. One node is common to all three, they intersect again by pairs in three nodes, and pass singly through nine others. The curves therefore pass through ten nodes obtained by excluding six coplanar nodes from the whole configuration.

We can exhibit these results clearly in a table showing the typical rationalising factors for the various equations and the number of nodes through which the curves of each family pass. We take $x_0x_{12}x_{13}x_{23}$ for a typical Rosenhain tetrad and suppose that x_0 is even and x_{12} is odd.

	even		odd	
$\Theta_0^{(2n)}$	1	0	$\sqrt{x_0x_{12}x_{13}x_{23}}$	16
$\Theta_{12}^{(2n)}$	$\sqrt{x_{13}x_{23}}$	8	$\sqrt{x_0x_{12}}$	8
$\Theta_0^{(2n+1)}$	$\sqrt{x_0}$	6	$\sqrt{x_{12}x_{13}x_{23}}$	10
$\Theta_{12}^{(2n+1)}$	$\sqrt{x_0x_{13}x_{23}}$	10	$\sqrt{x_{12}}$	6

§ 86. LINEAR SYSTEMS OF CURVES.

Each of the families of curves on Kummer's surface is a system of curves determined entirely by their order and base points, and is therefore a *complete linear system**. Now there are three important numbers connected with every linear system, namely its *dimension*, which is less by one than the number of linearly independent curves, its *degree*, which is the number of variable intersections of two curves, and the *deficiency* of the general curve of the system. All of these are unaltered by any birational transformation. We proceed to determine the first two by elementary methods.

The dimension of each system of curves is found by considering the number of conditions imposed on the corresponding family of surfaces at the base curves. In order that an m-ic surface may pass through a given conic it must be made to pass through $2m+1$ points of it, from which it easily follows that the

* See Castelnuovo et Enriques, *Math. Ann.* XLVIII, 241; Picard et Simart, *Fonctions algébriques de deux variables.*

numbers of conditions for an m-ic surface to pass through one, two, three or four conics of a Rosenhain tetrad are given by the following table.

conics	conditions
1	$2m+1$
2	$4m$
3	$6m-2$
4	$8m-4$

If in all cases the curves are of order $2n$ we get the following table which completes the preceding by giving the number of base

conics	order of surface	conditions	base nodes
1	$\frac{1}{2}(n+1)$	$n+2$	6
2	$\frac{1}{2}(n+2)$	$2n+4$	8
3	$\frac{1}{2}(n+3)$	$3n+7$	10
4	$\frac{1}{2}(n+4)$	$4n+12$	16
t	m	$tn+\frac{1}{2}(t^2+s)$	$2s$

nodes for the system of curves, so that the number of conditions in all cases may be expressed by the formula

$$tn+\tfrac{1}{2}(t^2+s).$$

Again so far as intersection with Φ is concerned the equation

$$S_m=0$$

is equivalent to $$S_m+S_{m-4}\Phi=0,$$

where S_{m-4} is an arbitrary polynomial of degree $m-4$. Hence by properly choosing S_{m-4} the number of arbitrary coefficients in the equation may be reduced to

$$\tfrac{1}{6}(m+1)(m+2)(m+3)-\tfrac{1}{6}(m-3)(m-2)(m-1)=2m^2+2.$$

In this put $m=\frac{1}{2}(n+t)$, subtract 1 for homogeneity, subtract also the preceding number of conditions, and we find for the dimension of the linear system of curves

$$\tfrac{1}{2}(n+t)^2+1-tn-\tfrac{1}{2}(t^2+s)=1+\tfrac{1}{2}(n^2-s).$$

We notice incidentally that the number of linearly independent $\Theta_{\alpha\beta}^{(n)}$, both even and odd, of any given characteristic, is n^2, for if an even $\Theta_{\alpha\beta}^{(n)}$ vanishes at $2s$ nodes, then an odd $\Theta_{\alpha\beta}^{(n)}$ vanishes at the remaining $16-2s$ nodes.

Next to find the degree of the system we recall that each curve is the curve of contact of a surface of order n, and any other curve of the system cuts this surface at $2s$ nodes and touches it at all the remaining points of meeting; hence the number of variable intersections is

$$\tfrac{1}{2}(2n^2-2s)=n^2-s.$$

The deficiency of a curve in space may be defined in various ways. It is possible in many ways to draw through the curve two surfaces of sufficiently high orders μ and ν, which will intersect again in one or more other curves. The surfaces of order $\mu+\nu-4$ passing through the residual intersection are called *adjoint* to the given curve, and although their definition leaves them to a great extent arbitrary, yet they cut the curve in a definite linear series of groups of points. If p is the deficiency of the curve, each of these groups consists of $2p-2$ points of which $p-1$ may be arbitrarily chosen. This series is called the *canonical series*. In the present case we put $\mu=4$, $\nu=\frac{1}{2}(n+t)$; the residual intersection consists of the t conics and the adjoint surfaces are the surfaces of the family. Hence the curves of the system cut any one of them in groups belonging to the canonical series, and on equating the two expressions for the number of points in each group

$$2p-2=n^2-s,$$

or

$$p=1+\tfrac{1}{2}(n^2-s),$$

so that in this case the deficiency is equal to the dimension.

CHAPTER XIV.

CURVES OF DIFFERENT ORDERS.

§ 87. QUARTIC CURVES.

After the sixteen singular conics, which are easily seen to illustrate the theorems and formulae of the preceding chapter, the simplest curves on the general Kummer surface are *quartics*, represented by equations of order two.

Taking first the characteristic to be zero, the equation, if even, represents the complete intersection with a surface of order one. We have then the family of plane sections, of which only four are linearly independent. In attempting to form an equation of zero characteristic we see from the multiplication table of the group that the two factors of each term must be equal, and the product is simply one of the sixteen linear forms. This shows that the equation cannot be odd, and therefore no quartic passes through all sixteen nodes.

Corresponding to any other characteristic there are two families of curves passing through complementary octads of nodes. Each family includes four pairs of conics and is cut out by a pencil of quadrics containing any one of the four pairs. The curves of the same family have no variable intersections since only one passes through an arbitrary point.

The curve cuts each trope of one octad at four nodes and each trope of the (complementary) octad at two nodes and therefore touches the latter at one point. Hence also the inscribed quadric touches each trope of an octad.

Two curves of associated families do not pass through any common nodes. Hence one *touches* the quadric inscribed along the other in four points, and the curves have four variable intersections. The two inscribed quadrics touch in four points and therefore cut in four generators.

Let $S_1=0$, $S_2=0$ be the quadrics inscribed along two curves of the same family, and let $S=0$ be the quadric containing both curves; then the equation of Kummer's surface can be written in the form

$$S_1S_2-S^2=0,$$

which is the envelope of the quadric

$$S_1+2\lambda S+\lambda^2S_2=0.$$

This surface touches Φ along any curve of the family.

If two fixed quadrics A, B have quadruple contact with a variable quadric C passing through a fixed point, the envelope of C is a Kummer surface. [Humbert, *Rendiconti di Circolo Matematico di Palermo*, XI, 1.]

At a common point of two quartics of associated families the two tangents are conjugate directions on the surface. [Darboux, *Comptes Rendus*, XCII, p. 1493.]

Two curves of families with different characteristics pass through the tetrad of nodes common to two octads. Hence of the eight intersections of one with the quadric inscribed along the other, four are at these nodes and the rest are at two points of contact. In this case there are *two* variable intersections and the inscribed quadrics touch at two points. From the incidence diagram we see that if two octads have a common tetrad, the two remaining tetrads together form an octad; accordingly a third family exists such that three curves, one from each family, cut by pairs in three tetrads of nodes. The three inscribed quadrics have double contact with each other. We shall now prove that the nature of the intersections of these quadrics depends upon whether the tetrads are even or odd.

It is necessary to form the equation of a family with given characteristic, say 12. Possible terms are

$$\sqrt{x_0x_{12}},\quad \sqrt{x_{34}x_{56}},\quad \sqrt{x_{35}x_{64}},\quad \sqrt{x_{36}x_{45}}$$

from one octad, and

$$\sqrt{x_{13}x_{23}},\quad \sqrt{x_{14}x_{24}},\quad \sqrt{x_{15}x_{25}},\quad \sqrt{x_{16}x_{26}}$$

from the associated octad. But since the equation of the surface is expressible as a linear relation between any three terms from each set of four, only two out of each set are linearly independent.

§ 88. QUARTICS THROUGH THE SAME EVEN TETRAD.

Select any four linear forms x, y, z, t forming an even or Göpel

$$\begin{matrix} t & x & \cdot & \cdot \\ y & z & \cdot & \cdot \\ \cdot & \cdot & \cdot & \cdot \\ \cdot & \cdot & \cdot & \cdot \end{matrix}$$

tetrad, as in the diagram (cf. p. 79); then the quartics

$$\sqrt{xt} + \lambda\sqrt{yz} = 0,$$

and

$$\sqrt{yt} + \mu\sqrt{zx} = 0,$$

pass through the four nodes represented by the same symbols as x, y, z, t, forming an even tetrad.

We know that $\sqrt{xyzt}$ is rational on Φ, so that the equation of Φ may be expressed in the form

$$xyzt = \phi^2,$$

and $\phi = 0$ is a quadric containing four conics. The equations of the inscribed quadrics are obtained, by squaring and rationalising, in the forms

$$A \equiv xt + 2\lambda\phi + \lambda^2 yz = 0,$$

$$B \equiv yt + 2\mu\phi + \mu^2 zx = 0,$$

whence

$$\mu A - \lambda B \equiv (\mu x - \lambda y)(t - \lambda\mu z),$$

showing that the complete intersection of A and B lies in two planes, and is therefore two conics.

There is a third family of quartics passing through the same four nodes, namely

$$\sqrt{zt} + \nu\sqrt{xy} = 0,$$

and the corresponding inscribed quadric is

$$C \equiv zt + 2\nu\phi + \nu^2 xy = 0.$$

Then

$$\nu B - \mu C \equiv (\nu y - \mu z)(t - \mu\nu x)$$

$$\lambda C - \nu A \equiv (\lambda z - \nu x)(t - \nu\lambda y)$$

$$\mu A - \lambda B \equiv (\mu x - \lambda y)(t - \lambda\mu z),$$

and the quadrics $B = 0$, $C = 0$ touch at two points on the line

$$\left.\begin{aligned} \nu y - \mu z &= 0 \\ t - \mu\nu x &= 0 \end{aligned}\right\}.$$

In this way we get three lines joining the pairs of points of contact of A, B, C, and they meet in the point

$$x/\lambda = y/\mu = z/\nu = t/\lambda\mu\nu;$$

each line cuts two opposite edges of the tetrahedron $xyzt = 0$.

The product

$$(\sqrt{xt}+\lambda\sqrt{yz})(\sqrt{yt}+\mu\sqrt{zx})(\sqrt{zt}+\nu\sqrt{xy})$$

is rational on Φ, whence we infer that a cubic surface $S=0$ can be found such that

$$ABC \equiv S^2 + G\Phi,$$

where G is a quadric. Evidently A and G touch along a conic lying on S; in fact

$$\begin{aligned} G &\equiv 4\mu\nu A - (t+\mu\nu x-\nu\lambda y-\lambda\mu z)^2 \\ &\equiv 4\nu\lambda B - (t-\mu\nu x+\nu\lambda y-\lambda\mu z)^2 \\ &\equiv 4\lambda\mu C - (t-\mu\nu x-\nu\lambda y+\lambda\mu z)^2. \end{aligned}$$

Hence A, B, C are all circumscribed about the same quadric and the three conics of contact and the three quartics (of contact with Φ) lie on a cubic surface.

The planes of contact of A and B with G intersect in the line joining the points of contact of A with B, proving once more that the three lines are concurrent.

Conversely, the three lines through an arbitrary point $(\lambda, \mu, \nu, \lambda\mu\nu)$ cutting pairs of opposite edges of an even tetrahedron of tropes, cut Φ in twelve points which include the six points of contact with a quadric G; the remaining six points are points of contact of a quadric G' obtained from G by changing the signs of λ, μ, ν. If A', B', C', are obtained from A, B, C in the same way, so that

$$A' \equiv xt - 2\lambda\phi + yz,$$

then
$$G - G' \equiv 4\mu\nu(A-A') = 4\lambda\phi,$$

and the pairs of quadrics G and G', A and A', etc. intersect in quartics lying on Φ.

Changing the signs of μ and ν merely permutes the quadrics so as to form the two sets $AB'C'$, $A'BC$. The points of contact of A and B', A' and B are collinear with $(\lambda, -\mu, -\nu, \lambda\mu\nu)$, and so on. In this way we obtain four points of concurrence forming a tetrahedron desmic with the tetrahedron of reference.

Thus a Göpel tetrahedron of tropes and any desmic tetrahedron determine six quadrics having twenty-four points of contact lying by fours on the six edges of the latter tetrahedron. The quadrics can be arranged in three pairs, each pair determining a pencil which includes the quadric passing through the conics of the Göpel tetrahedron. The eight sets of three quadrics, one from each pair, are circumscribed to eight quadrics, and the twenty-four conics of contact lie in the faces of the desmic tetrahedron.

§ 89. QUARTICS THROUGH THE SAME ODD TETRAD.

Consider now families of quartics passing through the same odd or Rosenhain tetrad of nodes (p. 78).

The last three rows in the diagram of incidences taken in pairs determine three octads of tropes and three families of quartics;

$$\begin{array}{cccc} \cdot & \cdot & \cdot & \cdot \\ \cdot & x & x' & \cdot \\ \cdot & y & y' & \cdot \\ \cdot & z & z' & \cdot \end{array}$$

any two of the octads contain a common Rosenhain tetrad of nodes through which the corresponding families pass. Now since each family is of only one dimension we may take the equations to be

$$\sqrt{yz} + \lambda \sqrt{y'z'} = 0,$$
$$\sqrt{zx} + \mu \sqrt{z'x'} = 0,$$
$$\sqrt{xy} + \nu \sqrt{x'y'} = 0.$$

Further $\sqrt{xx'}$, $\sqrt{yy'}$, $\sqrt{zz'}$ are linearly connected in virtue of $\Phi = 0$, and we may take the irrational equation of the surface to be

$$\sqrt{xx'} + \sqrt{yy'} + \sqrt{zz'} = 0.$$

Then, by squaring and rationalising, the three inscribed quadrics are

$$\begin{aligned} A &\equiv yz + \lambda (xx' - yy' - zz') + \lambda^2 y'z' \\ &\equiv (y - \lambda z')(z - \lambda y') + \lambda x x' = 0, \\ B &\equiv (z - \mu x')(x - \mu z') + \mu y y' = 0, \\ C &\equiv (x - \nu y')(y - \nu x') + \nu z z' = 0. \end{aligned}$$

Hence B and C contain the line

$$\left.\begin{aligned} x - \nu y' - \mu z' &= 0 \\ \mu y + \nu z - \mu\nu x' &= 0 \end{aligned}\right\}$$

and have no other common generator. Hence the remaining intersection is a cubic curve. Evidently the three common generators lie in the plane

$$\lambda x + \mu y + \nu z = \mu\nu x' + \nu\lambda y' + \lambda\mu z'.$$

Since the line common to B and C must touch Φ where it meets it, the two variable points of intersection of the quartics are the points of contact of a bitangent to Φ.

As before, the product

$$(\sqrt{yz} + \lambda \sqrt{y'z'})(\sqrt{zx} + \mu \sqrt{z'x'})(\sqrt{xy} + \nu \sqrt{x'y'})$$

is rational on Φ, and therefore the three quartics form the complete intersection with a cubic surface S, and the three quadrics are circumscribed about a quadric G in consequence of the identity

$$ABC \equiv S^2 + G\Phi.$$

Since S contains the curves (A, Φ), (B, Φ), it touches Φ at their points of intersection. Hence the common generator of A and B is a bitangent of the cubic surface S and therefore lies entirely on it. Since the three lines common to (AB), (BC), (CA) lie on S, they must also lie on G which therefore consists of their plane repeated.

Since through any point of Φ six bitangents can be drawn, and there are thirty octads, it follows that five quartics from different families cut in the same two points. It is easy to see that the five characteristics together with the zero characteristic make up a set representing six coplanar nodes.

§ 90. SEXTICS THROUGH SIX NODES.

There are thirty-two families of sextic curves on the surface, two for each characteristic. They are of two kinds: sixteen are cut out by surfaces of order $\frac{1}{2}(n+1), = 2$, passing through a conic, and the other sixteen are cut out by surfaces of order $\frac{1}{2}(n+3), = 3$, passing through three concurrent conics.

Taking a family of the first kind, the quadric is subjected to five conditions in containing a given conic, and there remain five arbitrary coefficients. Hence the family contains *five* linearly independent curves.

The sextic meets one trope at six nodes and every one of the fifteen others at two nodes and two points of contact. The inscribed cubic surface cuts the one trope in a plane cubic passing through the six nodes, and every other trope in a conic and the line joining the two points of contact.

Two quadrics through the same conic cut again in another conic; hence two sextics of the same family have six variable intersections lying on a conic. The corresponding inscribed cubic surfaces touch at these six coplanar points and therefore the plane cuts them in the same cubic curve.

Among the quadrics cutting out the family there are fifteen containing the four conics of a Göpel tetrad, for when one conic is given, three others can be found in fifteen ways to complete a Göpel tetrad, and all four lie on the same quadric. Hence fifteen sextics of the family break up into three conics, and in each case the inscribed cubic surface breaks up into the three tropes containing the conics. By the last paragraph these three conics cut any sextic of the family in six coplanar points, other than the six common nodes, and the three lines in which this plane cuts the three tropes lie also on the inscribed cubic surface, which has thus been shown *to contain fifteen lines, one in each of fifteen tropes, lying by threes in fifteen tritangent planes.*

If $S_1 = 0$, $S_2 = 0$, are the cubics inscribed along two sextics of the same family, and $S = 0$ is the cubic containing both, we have the identity

$$S_1 S_2 \equiv S^2 + G\Phi$$

where G is a quadric. Now we have seen that S_1 and S_2 touch one another at the six variable intersections of the sextics and therefore also touch S and Φ. Hence S contains the common plane cubic section of S_1 and S_2, and G must be this plane repeated.

We proceed to demonstrate these theorems analytically.

Let x, y, z, t, be a Göpel tetrad of linear forms; then, since $xyzt$ is an even product of characteristic zero, the five products xyz, x^2t, y^2t, z^2t, t^3 are linearly independent and of the same order, characteristic and parity. The equation of the family of sextics passing through the six nodes in the trope $t = 0$ is

$$\sqrt{xyz} + (ax + by + cz + dt)\sqrt{t} = 0.$$

Write the equation of Kummer's surface in the form

$$\Phi \equiv xyzt - \phi^2 = 0$$

so that $\phi = 0$ is the quadric containing the tetrad of conics. Then the linear system of ∞^4 quadrics cutting out the family is represented by the equation

$$\phi + (ax + by + cz + dt)\,t = 0.$$

By squaring and rationalising using $\Phi = 0$ we find the inscribed cubic to be

$$S_1 \equiv xyz + 2(ax + by + cz + dt)\,\phi + (ax + by + cz + dt)^2\,t = 0.$$

Obviously one tritangent plane is

$$ax + by + cz + dt = 0.$$

Another sextic of the family is

$$\sqrt{xyz} + (a'x + b'y + c'z + d't)\sqrt{t} = 0,$$

and the common points lie on the curve

$$(\overline{a-a'}x+\overline{b-b'}y+\overline{c-c'}z+\overline{d-d'}t)\sqrt{t}=0$$

consisting of a conic and a plane section of Φ.

Write for abbreviation

$$u_1 \equiv ax+by+cz+dt$$
$$u_2 \equiv a'x+b'y+c'z+d't,$$

then

$$S_1 \equiv xyz+2u_1\phi+u_1^2t$$
$$S_2 \equiv xyz+2u_2\phi+u_2^2t$$
$$S \equiv xyz+(u_1+u_2)\phi+u_1u_2t,$$

whence

$$S_1S_2-S^2 \equiv (u_1-u_2)^2(xyzt-\phi^2)$$

and

$$S_1-S \equiv (u_1-u_2)(\phi+u_1t)$$
$$S_2-S \equiv (u_2-u_1)(\phi+u_2t),$$

showing that the three cubics have a common curve in the plane $u_1=u_2$.

Two sextics from different families cut in two nodes and eight other points. We may take their equations to be

$$\sqrt{yzt}+u_1\sqrt{x}=0,$$
$$\sqrt{xzt}+u_2\sqrt{y}=0.$$

They both lie on the quartic surface obtained by rationalising

$$\sqrt{zt}\,(\sqrt{yzt}+u_1\sqrt{x})(\sqrt{xzt}+u_2\sqrt{y})=0,$$

that is, on

$$zt(\phi+u_1x+u_2y)+u_1u_2\phi=0.$$

If in particular we take $u_1\equiv u_2\equiv u$, the points of intersection lie on the two quadrics

$$\phi+ux=0,$$
$$\phi+uy=0,$$

cutting in two conics in the planes $u(x-y)=0$. The conic in $x-y=0$ cuts Φ in the two common nodes and four other common points. The conic in $u=0$ lies on $\phi=0$, and cuts Φ in four pairs of points on the four conics in $xyzt=0$. Hence the remaining four common points are the points of contact of the lines $u=0=z$, and $u=0=t$ with Φ.

We are thus led to consider the four sextics

$$\alpha\sqrt{yzt}+u\sqrt{x}=0,$$
$$\beta\sqrt{ztx}+u\sqrt{y}=0,$$
$$\gamma\sqrt{txy}+u\sqrt{z}=0,$$
$$\delta\sqrt{xyz}+u\sqrt{t}=0,$$

cut out by four quadrics, one through each conic of a Göpel tetrad, and the same plane section of the quadric containing the four conics. They pass by threes through the pairs of points $u = 0$, $x = 0$, $\phi = 0$, etc., and the remaining twenty-four intersections lie by fours on the six concurrent planes

$$x/\alpha = y/\beta = z/\gamma = t/\delta.$$

§ 91. SEXTICS THROUGH TEN NODES.

The sextics of a family of the second kind are cut out by cubic surfaces passing through three concurrent conics. The number of conditions for the cubic is sixteen, leaving *four* linearly independent surfaces.

The sextic passes through the ten nodes not lying on the trope which completes the Rosenhain tetrad, and therefore has three contacts with this plane and one contact with every other trope. The same is true of the inscribed cubic which therefore contains three lines lying in one trope.

Two curves of the same family cut at ten nodes and at four other points. The two inscribed cubics S_1, S_2, and the cubic S containing both curves, all touch Φ at the same four points, and further we have the identity

$$S_1 S_2 - S^2 \equiv G\Phi,$$

where G is a quadric. This shows that the twenty-seven common points of the three cubics are singular points on $G\Phi = 0$. Now the four contacts count for sixteen intersections and ten more are at ten nodes of Φ; there remains one not lying on Φ which must be a node on G. Hence *G is a cone* circumscribing S_1 and S_2 along the residual cubic intersections with S.

Again points common to S, G, Φ are singular points on $S_1 S_2 = 0$. These include the four points of contact of S_1 and S_2 with Φ, each counted four times, and four more on each sextic which are therefore nodes on S_1 and S_2.

Since the inscribed cubic has four nodes the lines joining them lie entirely on it and therefore touch Φ where they cut the sextic of contact. Hence *the four nodes are the corners of a tetrahedron inscribed in Φ, whose edges touch Φ.* The six edges are torsal lines on the nodal cubic surface and the six pinch planes cut the surface in three lines cutting the pairs of non-intersecting torsal lines and lying in a tritangent plane. The cubic possesses only one tritangent plane beside those containing the nodes, and in the present case this has been shown to be one of the tropes of Φ*.

In order to construct the nodes of an inscribed cubic surface we have only to take any three points on one conic of Φ and draw the six tangent planes through pairs of them. These planes meet by threes in the four points required. Thus an inscribed tetrahedron whose edges touch Φ can be constructed in ∞^3 ways†.

§ 92. OCTAVIC CURVES THROUGH EIGHT NODES.

There are thirty-two families of three different kinds. The first is the family of complete intersections with quadric surfaces, and is of no particular interest. It is represented by an even equation of order four and zero characteristic, containing ten terms.

Corresponding to each of the thirty octads is a family of octavics cut out by cubic surfaces passing through any one of the four pairs of conics containing the octad. There are eight terms in the equation. Four linearly independent curves of the family are given by a pair of conics and any plane section; let $xt = 0$ and $yz = 0$ be two pairs of planes containing the octad, then the equation of the family can be written in the form

$$(ax + by + cz + dt)\sqrt{xt} + (a'x + b'y + c'z + d't)\sqrt{yz} = 0,$$

or

$$u\sqrt{xt} + v\sqrt{yz} = 0.$$

Evidently the four points in which the line $u = 0 = v$ meets Φ lie on this curve, so that the line is a *quadruple secant*.

The octavic passes through an octad of nodes and therefore meets each of an octad of tropes at four nodes and two contacts, and each trope of the associated octad at two nodes and three contacts. The inscribed quartic surface has eight tropes of Φ for tritangent planes.

The planes $xyzt = 0$ form a Göpel tetrahedron of tropes and so the equation of Φ may be taken to be

$$xyzt = \phi^2$$

and then the inscribed quartic is

$$u^2xt + 2uv\phi + v^2yz = 0$$

having the quadruple secant for a double line; there is a pencil of quartics

$$xyzt - \phi^2 + \lambda(u^2xt + 2uv\phi + v^2yz) = 0$$

* These theorems are easily proved by taking the tetrahedron of nodes for reference.

† Humbert, *Liouville*, sér. 4, IX, 103.

all touching Φ along the same curve. By writing this equation in the form

$$(yz + \lambda u^2)(xz + \lambda v^2) - (\phi - \lambda uv)^2 = 0$$

we see that the surface has eight nodes, common to three quadrics; for $\lambda = \infty$ the nodes coincide by pairs, at the pinch points on the nodal line $u = 0 = v$.

The fact that the inscribed quartics have eight nodes on the octavic curve of contact can be inferred from the identity

$$S_1 S_2 \equiv S^2 + G\Phi,$$

for each of the twelve common points of contact of S_1, S_2, Φ is counted four times among the points common to S, G, Φ, leaving sixteen singular points of $S^2 + G\Phi = 0$ to be divided between S_1 and S_2.

§ 93. OCTAVIC CURVES THROUGH SIXTEEN NODES.

The remaining family is represented by an odd equation of order four and zero characteristic, and the curves are cut out by quartic surfaces through an odd tetrad of conics. The surfaces must be made to pass through the four corners of the tetrad and six more points on each conic, making twenty-eight conditions; now a quartic surface contains thirty-five terms, but for purposes of intersection these terms are connected by one linear relation $\Phi = 0$. Hence the family contains *six* linearly independent curves. We have already had examples of these curves in the principal asymptotic curves (p. 62) and in the curves of contact of inscribed Kummer surfaces (p. 66).

Two curves of the family cut in eight points other than nodes; at these points the inscribed quartics touch. The identity

$$S_1 S_2 \equiv S^2 + G\Phi$$

shows that the sixty-four points common to S_1, S_2, S are singular points on $G\Phi = 0$. Of these, thirty-two are accounted for by the intersections of the octavic curves and sixteen are at the nodes of Φ; hence the remaining sixteen are nodes of G, which is therefore a Kummer surface. Similar reasoning shows that S_1 and S_2 have also sixteen nodes each. Hence the quartics touching a given Kummer surface along an octavic curve passing through all the nodes, are also Kummer surfaces. Of these inscribed quartic surfaces ten have double lines and are Plücker surfaces (p. 68). Hence the octavic has ten quadruple secants.

CHAPTER XV.

WEDDLE'S SURFACE.

§ 94. BIRATIONAL TRANSFORMATION OF SURFACES.

Let P, Q, R, S be any four homogeneous polynomials in x, y, z, t, of the same degree. The equations

$$X/P = Y/Q = Z/R = T/S$$

establish a correspondence between two spaces: to any point (x, y, z, t) of one space corresponds one point (X, Y, Z, T) of the other, and to the latter correspond the n variable points common to the three surfaces

$$P/X = Q/Y = R/Z = S/T.$$

A special case occurs when $n = 1$. The correspondence is then unique and therefore rational both ways, and the equations constitute a rational transformation between two spaces*.

If (x, y, z, t) describes a surface

$$f(x, y, z, t) = 0,$$

the corresponding point (X, Y, Z, T) describes a surface

$$F(X, Y, Z, T) = 0$$

into which f is transformed. The plane sections

$$aX + bY + cZ + dT = 0$$

correspond to the *linear system* of curves cut out on f by the family of surfaces

$$aP + bQ + cR + dS = 0.$$

In general those curves of this system which pass through an assigned point do not all pass through another point: accordingly to a point (X, Y, Z, T) on F determined by three planes corresponds on f the unique point common to the three corresponding curves of the linear system. In other words the correspondence

* Cayley, VII, 189.

between the two surfaces is unique, and therefore by means of the equation $f=0$ the equations of transformation can be solved for x, y, z, t rationally in terms of X, Y, Z, T. For this reason the transformation between the two surfaces is called *birational.*

If a simple point of f is a multiple base point, we have approximately

$$X/P_m = Y/Q_m = Z/R_m = T/S_m,$$

and the denominators can be expressed as rational m-ic polynomials in y/x; hence the base point is transformed into a rational m-ic curve. More generally, if the point has multiplicity m' on f, the corresponding curve is of order mm'. For example, a node on f at which the tangent cone is expressible parametrically in the form $x = y/\theta = z/\theta^2$ is transformed into a rational $2m$-ic curve. An important exceptional case is when the cones $P_m = 0$, $Q_m = 0$, $R_m = 0$, $S_m = 0$ have a common generator; taking this to be $x = 0$, $y = 0$, we find that after substitution the highest power of θ is θ^{2m-1}, and in this case the node is transformed into a rational $(2m-1)$-ic curve. Thus, for example, a node through which all the surfaces of the family pass is in general transformed into a conic, but if it lies on a simple base curve it is transformed into a straight line.

Next consider a simple curve on f which is a simple base curve for the family of surfaces. At any point of it, the tangent line being $x = 0 = y$, P_1, Q_1, R_1, S_1 are linear in x and y only, and for near points on f, y/x has one value, so that the ratios $P_1 : Q_1 : R_1 : S_1$ are definite. There is therefore a unique corresponding point on F, and its locus represents the base curve. More generally, if a m-ple base curve is m'-ple on f, y/x has m' values giving m' points on F lying on a rational m-ic. The locus of these points is a simple curve on F corresponding to the multiple curve on f.

If only ∞^2 surfaces of the family pass through a simple curve on f, we may take three of them to be P, Q, R: then the whole curve is transformed into the single point $X = Y = Z = 0$. The multiplicity of this point is equal to the number of ratios $X : Y : Z$ satisfying

$$aX + bY + cZ = 0,$$

corresponding to points (x, y, z) in the neighbourhood of the curve, and this is equal to the number of points in which the curve cuts the residual intersection of $f = 0$ with

$$aP + bQ + cR = 0.$$

The transformation depends on the linear system of curves and not on the particular surfaces cutting them out, for these may be modified by means of the equation $f=0$. The *degree* or number of variable intersections of two curves of the system is equal to the number of intersections of two plane sections of the new surface, that is, its *order*, and further, for the transformation to be possible the *dimension* or multiplicity of the system must be at least three.

Let the points $x=y=z=0$, $X=Y=Z=0$ correspond on the two surfaces f and F. Put $t=1$, $T=1$, and let P, Q, R, S be expanded in series of homogeneous polynomials in x, y, z, of degrees indicated by suffixes, thus

$$P = P_1 + P_2 + \ldots,$$
$$Q = Q_1 + Q_2 + \ldots,$$
$$R = R_1 + R_2 + \ldots,$$
$$S = S_0 + S_1 + S_2 + \ldots,$$

since (0, 0, 0) is not supposed to be a base point. From this it is obvious that the tangent cone at a multiple point on f is linearly transformed by the equations

$$X/P_1 = Y/Q_1 = Z/R_1 = 1/S_0$$

into the tangent cone at the corresponding point on F, so that *corresponding points have the same multiplicity.*

Next, let (0, 0, 0) be a simple base point, so that $S_0 = 0$. We now have approximately

$$X/P_1 = Y/Q_1 = Z/R_1 = 1/S_1,$$

and by means of the equation of the tangent plane to f the four denominators can be expressed as linear functions of x and y. Hence to the pencil of tangent lines to f correspond the points of a *straight line* on F, that is, a rational curve of order 1, as was shown before.

§ 95. TRANSFORMATION OF KUMMER'S SURFACE.

Among the birational transformations of a given surface the most useful are those in which the order of the new surface is as low as possible, and also the order of the surfaces employed in the transformation. We require a linear system of curves having as many intersections as possible at base points: so we shall consider only those systems which are contained in the families of curves already investigated, the number of variable intersections being further diminished by fixed multiple base points*.

The base points of the system may be of three kinds, according as they are at base nodes of the family, or at other nodes, or at ordinary points. In the first case, if the curve be cut out by a surface passing through $2t-1$ concurrent ($t=1, 2, 3$) conics and having a $(\lambda+t)$-ple point, the multiplicity of the point on the curve will be $2\lambda+1$. To find the number of coincident intersections of two such curves we employ the principle of continuity, and vary the two surfaces until each tangent cone breaks up into planes; $t-1$ of these planes may be supposed to contain the tangent lines to $2t-2$ conics, and may be thrown off. We are left with $\lambda+1$ planes through the node of which one passes through the tangent line to a conic. Hence the number of coincident intersections of the two curves is

$$2\lambda^2+2\lambda+1,$$

and the consequent reduction of degree is $2\lambda(\lambda+1)$.

In the second case, the curve must have a point of even multiplicity 2μ, the surface cutting it out having a μ-ple point. The number of coincident intersections of two curves of the system is the number of coincident intersections of three cones of orders $\mu, \mu, 2$, that is, $2\mu^2$, and this is the corresponding reduction of degree. Thirdly, a ν-ple point at an ordinary point gives ν^2 coincident intersections.

Now if the curves of the family are of order $2n$ and pass through $2s$ base nodes, the degree of the family is n^2-s (p. 148); hence the degree of the new system is

$$N=n^2-s-\sum^{2s}(2\lambda^2+2\lambda)-\sum^{16-2s}2\mu^2-\Sigma\nu^2,$$

so that $$2N=2n^2-\sum^{2s}(2\lambda+1)^2-\sum^{16-2s}(2\mu)^2-2\Sigma\nu^2.$$

A birational transformation of Kummer's surface is effected by means of four linearly independent curves of this system, and is therefore possible only if the dimension is sufficiently great. It is necessary to determine the number of conditions that curves of a given family may have assigned singularities. In the first case, since the $2s$ base nodes may be treated alike, we require the number of conditions that the curve cut out on Kummer's surface $\Phi=0$, by a surface $F=0$, may have a $(2\lambda+1)$-ple point

* Cf. Humbert, *Liouville*, sér. 4, IX, 449, who obtains the same results by transcendental methods. See also Hutchinson, *Amer. Bull.* VII, 211.

at a node on a conic through which F passes. Take this node for origin and the tangent to the conic for axis of z. If F has a k-ple point the number of conditions for a $(k+1)$-ple point is

$$\tfrac{1}{2}(k+1)(k+2)-1,$$

since the term z^k is absent. Hence when F is arbitrary, subject only to the condition of passing through the conic, the number of conditions for a $(\lambda+1)$-ple point is

$$\tfrac{1}{6}(\lambda+1)(\lambda+2)(\lambda+3)-(\lambda+1).$$

But as far as the curve of intersection is concerned the surface $F=0$ may be replaced by $F+G\Phi=0$, and the arbitrary coefficients in G of the terms of order $\leqslant \lambda-2$ may be used to satisfy

$$\tfrac{1}{6}(\lambda-1)\lambda(\lambda+1)$$

of these conditions. We are left with $\lambda(\lambda+1)$ conditions among the coefficients in F alone.

In the second case, the terms of order $\leqslant \mu-1$ in $F+G\Phi$ must disappear at a node of Φ, and for this the terms of G of order $\leqslant \mu-3$ may be used, leaving

$$\tfrac{1}{6}\mu(\mu+1)(\mu+2)-\tfrac{1}{6}(\mu-2)(\mu-1)\mu=\mu^2$$

conditions among the coefficients in F alone.

In the third case, the terms of order $\leqslant \nu-1$ in $F+G\Phi$ must disappear at a simple point of Φ, and for this the terms of G of order $\leqslant \nu-2$ may be used; the number of conditions for F is therefore

$$\tfrac{1}{6}\nu(\nu+1)(\nu+2)-\tfrac{1}{6}(\nu-1)\nu(\nu+1),\ =\tfrac{1}{2}\nu(\nu+1).$$

Now (p. 148) the dimension of the family is $1+\frac{1}{2}(n^2-s)$, and therefore the dimension of the new linear system determined by the preceding three kinds of base points is

$$D=1+\tfrac{1}{2}(n^2-s)-\sum^{2s}\lambda(\lambda+1)-\sum^{16-2s}\mu^2-\Sigma\tfrac{1}{2}\nu(\nu+1).$$

Hence $$N-2D=-2+\Sigma\nu.$$

Now for the transformation to be possible we must have

$$D\geqslant 3,$$

and therefore $$N\geqslant 4+\Sigma\nu.$$

§ 96. QUARTIC SURFACES INTO WHICH KUMMER'S SURFACE CAN BE TRANSFORMED.

The last result shows that when $N = 4$, $\Sigma\nu = 0$, and therefore every $\nu = 0$, that is, *for a birational transformation* (of Kummer's surface) *into another quartic surface, all the base points must be at nodes**. Further, the multiplicities at the base points must be (p. 163) so chosen that

$$\sum^{2s} (2\lambda + 1)^2 + \sum^{16-2s} (2\mu)^2 = 2n^2 - 8,$$

and then the dimension of the system is exactly 3, so that in this case *the linear system of curves by which the transformation is effected is complete*, for it is determined entirely by its base points.

We now attempt to satisfy this equation for different values of n and s, that is, we have to express $2n^2 - 8$ as the sum of $2s$ odd squares and $16 - 2s$ even squares.

Taking $n = 2$ we must have $s = 0$, $\mu = 0$, and we get the general linear transformation.

Taking $n = 3$, $s = 3$ the only possible way is

$$1 + 1 + 1 + 1 + 1 + 1 + 4 = 10,$$

and the transformation is effected by sextic curves passing through six coplanar nodes and having a double point at one other node; these are cut out by quadrics through one conic and a seventh node. This is projectively equivalent to inversion†.

Next, taking $n = 3$, $s = 5$, $2(n^2 - 4) = 10$. The only way is

$$1 + 1 + 1 + 1 + 1 + 1 + 1 + 1 + 1 + 1 = 10,$$

and the transformation is effected by sextic curves passing through ten nodes and cut out by cubic surfaces through three concurrent conics. This leads to the surface which is considered in the next section. It may be remarked that this is the only case in which the family of curves receives no additional base points, for the condition for this is

$$2s = 2(n^2 - 4),$$

or

$$n^2 = s + 4,$$

* This theorem follows at once from the fact that the rational curve into which a base point, not at a node, is transformed, lies on an "adjoint" surface of order $N - 4$.

† Nöther, *Math. Ann.* III, 557; Cayley, *Proc. London Math. Soc.* III, 170; *Coll. Papers*, VII, 230.

and since s can have only the values 0, 1, 3, 4, 5, 8, the only admissible value is $s=5$, giving $n=3$.

Any two of the cubic surfaces cut in three conics and a variable cubic curve passing through the point of concurrence of the conics. This curve cuts each conic in two other points and therefore of its nine intersections with a third cubic surface, eight are on the base curves, leaving *one* variable intersection.

This is a particular case of the lineo-linear transformation considered by Cayley (*Coll. Papers*, VII, 236) and Nöther (*Math. Ann.* III, 517).

§ 97. WEDDLE'S SURFACE*.

We now consider in detail the surface W into which Kummer's surface Φ is transformed by means of sextic curves passing through ten nodes (cf. p. 157). These will be termed 'even' nodes (123.456), etc. as distinguished from the remaining six 1, 2, 3, 4, 5, 6 which are 'odd,' and lie in the trope x_0, in the notation of pp. 16, 18, 140.

By the general theory of transformation, the six odd nodes, not being base points, become nodes on the new surface W, while the ten even nodes become straight lines.

The trope x_0 meets every sextic of the family at *three* points of contact. Hence the conic in x_0 is transformed into a curve which cuts every plane section of W in three points. Since it passes through the six nodes of W it must be the unique cubic curve determined by them.

Any other trope, x_{12}, meets every sextic at four nodes and *one* point of contact. Hence the conic in x_{12} is transformed into a straight line joining the two nodes on W corresponding to the two nodes common to x_0 and x_{12}.

Through each of the ten even nodes (123.456) pass two sets of three tropes x_{23}, x_{31}, x_{12} and x_{56}, x_{64}, x_{45}, each of which forms with x_0 a Rosenhain tetrad. The two sets of three conics are reducible sextics of the family and correspond to the plane sections of W through the nodes 1, 2, 3 and 4, 5, 6 respectively. The node (123.456) is transformed into a line which must lie on both of these planes. Hence W contains the ten lines of intersection of two planes containing all six nodes, and these lines correspond to the ten even nodes of Kummer's surface.

* First mentioned by T. Weddle, *Camb. and Dubl. Math. Jour.* (1850), v, 69, note.

We have now established the following correspondence between the two surfaces.

Φ Kummer's surface	W Weddle's surface
six odd nodes	six nodes
ten even nodes	ten lines in planes of three nodes
fifteen conics	fifteen joins of nodes
one conic	cubic through nodes
sextic through even nodes	plane section
plane section	sextic

Let p, q, r, s be four products of three linear forms $x_{12}\dots$ such that each completes with x_0 a Rosenhain tetrad and such that $\sqrt{p}, \sqrt{q}, \sqrt{r}, \sqrt{s}$ are linearly independent on Φ. Then the equations of transformation may be taken to be (cf. p. 143)

$$X/\sqrt{p} = Y/\sqrt{q} = Z/\sqrt{r} = T/\sqrt{s}.$$

For example if the matrix of linear forms (p. 29) is

$$\begin{pmatrix} x_0 & x & y & z \\ . & x' & y' & z' \\ . & . & . & . \\ . & . & . & . \end{pmatrix}$$

we may take

$$p = xy'z', \quad q = x'yz', \quad r = x'y'z, \quad s = xyz.$$

The section of W by a quadric surface

$$F(X, Y, Z, T) = 0$$

corresponds to the curve on Φ given by

$$F(p^{\frac{1}{2}}, q^{\frac{1}{2}}, r^{\frac{1}{2}}, s^{\frac{1}{2}}) = 0,$$

an equation of order six, zero characteristic, and even, representing a curve of order twelve, the complete intersection of a cubic surface passing through the ten even nodes. If the quadric $F(X, Y, Z, T) = 0$ passes through any of the nodes of W, the cubic surface passes through the corresponding odd nodes of Φ. Hence a quadric through all the six nodes of W corresponds to a cubic through all the sixteen nodes of Φ, that is, to the polar surface of any point, and therefore to a plane section of the reciprocal surface. Now the reciprocal of a Kummer surface Φ is another Kummer surface Φ', and so we have deduced a birational transformation between Φ' and W by means of quadric surfaces through the six nodes of W*.

In order to reverse this transformation we require the curves on Φ' corresponding to the sextics on Φ, that is the reciprocals of the developables circumscribed to Φ along a sextic. The enveloping cone from any point touches Φ along a curve of order twelve, passing through all the nodes and cutting any one of the sextics, along which a cubic surface can be inscribed, in $\frac{1}{2}3.12=18$ points. Of these ten are at nodes, leaving eight variable intersections. Hence the class of the developable is eight. Of the tropes, x_0 touches the sextic at three points and is therefore a triple plane of the developable; every other trope touches the sextic once. Hence the corresponding curve on Φ' is an octavic which passes through all the nodes, and has a triple point at one of them. This is the case of birational transformation when $n=4$, $s=8$; the number $2n^2-8, =24$, must be expressed as the sum of sixteen odd squares, and this can be done in only one way

$$1+1+\ldots+1+9=24.$$

The surfaces cutting out these curves are quartics through a Rosenhain tetrad of conics having a node at one of the nodes of Φ, other than a corner of the tetrahedron. It may easily be verified that three surfaces of the system cut in only one arbitrary point.

We do not give an independent investigation but give references to the literature of the subject. We are concerned with the surface as a birational transformation of Kummer's surface, and in this view all its properties may be deduced from known properties of Kummer's surface. Attention must be called to the correspondence between the sheaves of lines through the nodes of W and the six quadratic congruences of bitangents on Φ, which may be explained as follows.

Two quadrics through the six nodes of W cut in a quartic curve which cuts W in sixteen points of which twelve are counted at the nodes. The remaining four points correspond to the four points of intersection of Φ with the line common to the planes corresponding to the two quadrics. Since tangent planes to Φ correspond to cones through the six nodes of W, a bitangent to Φ corresponds to the intersection of two cones, each of which passes through the vertex of the other. Their quartic intersection must therefore break up into the line joining their vertices and a twisted

* For this transformation consult De Paolis, *Memoire Lincei*, ser. 4, I, 576; *Rendiconti Lincei*, ser. 4, VI, 3.

cubic. Since the complete intersection passes through all the nodes, the cubic must pass through five of them, and the straight line through the sixth. Hence any two points on W collinear with a node correspond to the points of contact of a bitangent to Φ. The theorems already proved for bitangents can be at once carried over, for example:

Complete quadrilaterals can be inscribed in W having two opposite corners at nodes.

By successively projecting any point of W from the nodes, on to W, and also the projections, a system of only 32 points is obtained*.

§ 98. EQUATION OF WEDDLE'S SURFACE.

Let

$$P \equiv \Sigma P_{rs} x_r x_s = 0$$

$$Q \equiv \Sigma Q_{rs} x_r x_s = 0 \qquad (r, s = 1, 2, 3, 4)$$

$$R \equiv \Sigma R_{rs} x_r x_s = 0 \qquad (P_{rs} = P_{sr} \text{ etc.})$$

$$S \equiv \Sigma S_{rs} x_r x_s = 0$$

be four linearly independent quadrics passing through six points; then every other quadric through the same points is obtained by linearly combining these in the form

$$aP + bQ + cR + dS = 0.$$

Any surface

$$W(x_1, x_2, x_3, x_4) = 0$$

is transformed, by taking P, Q, R, S as new point coordinates, into a surface

$$\Psi(P, Q, R, S) = 0.$$

To any point (P, Q, R, S) on Ψ corresponds a point (x_1, x_2, x_3, x_4) or briefly (x) on W, and another point (x'), forming with (x) and the six base points a group of eight associated points. The locus of (x') is another surface $W' = 0$, and there must be an identity of the form

$$\Psi(P, Q, R, S) \equiv WW'$$

after both sides have been expressed in terms of x_1, x_2, x_3, x_4.

The locus of a point (x) which coincides with the eighth associated point (x') is the Jacobian surface

$$J \equiv \frac{\partial(P, Q, R, S)}{\partial(x_1, x_2, x_3, x_4)} = 0.$$

* Baker, *Proc. Lond. Math. Soc.*, ser. 2, I, 247.

Now $J=0$ is the quartic surface* which is the locus of the vertices of cones passing through the six base points of the family

$$aP+bQ+cR+dS=0,$$

for the conditions for a cone with vertex at (x_1, x_2, x_3, x_4) are

$$a\partial P/\partial x_s+b\partial Q/\partial x_s+c\partial R/\partial x_s+d\partial S/\partial x_s=0 \quad (s=1, 2, 3, 4)$$

and on eliminating a, b, c, d we get $J=0$, and from this it follows at once that J contains the fifteen joins of the six points and the ten lines of intersection of two planes containing all the points. If W is the surface considered in the preceding section, J and W intersect in twenty-five lines, and therefore coincide; and then W', which passes through all points common to J and W, also coincides with J. Ψ is then the surface Φ' reciprocal to the Kummer surface first considered and we have the important identity†

$$\Phi'(P, Q, R, S)\equiv J^2$$

which admits of direct verification.

Let four of the base points be taken for tetrahedron of reference and let the remaining two be (e_1, e_2, e_3, e_4) and (f_1, f_2, f_3, f_4). Then the equation of Weddle's surface can be expressed in the convenient form‡:

$$\begin{vmatrix} e_1f_1x_1^{-1} & x_1 & e_1 & f_1 \\ e_2f_2x_2^{-1} & x_2 & e_2 & f_2 \\ e_3f_3x_3^{-1} & x_3 & e_3 & f_3 \\ e_4f_4x_4^{-1} & x_4 & e_4 & f_4 \end{vmatrix}=0$$

which is the same as $\Sigma\, e_sf_s\xi_s/x_s=0$ where $\xi_s=0$ are the four planes through the line ef and the corners of reference.

The values of $a:b:c:d$ derived from the four equations

$$\frac{\partial}{\partial x_s}(aP+bQ+cR+dS)=0 \qquad (s=1, 2, 3, 4)$$

are the coordinates of a plane section of Φ' which corresponds to a cone of the family of quadrics, that is, of a tangent plane of Φ'. Hence the result of eliminating x_1, x_2, x_3, x_4 is to give the

* Darboux, *Bull. des Sciences math.* (1870), I, 348; Caspary, *ibid.* (1887), XI, 222; Hierholzer, *Math. Ann.* IV, 172.

† Proved otherwise by Schottky by a beautiful piece of reasoning, *Crelle*, CV, 238.

‡ Caspary, *Comptes Rendus*, CXII, 1357. Hutchinson, *Annals of Mathematics*, XI, 158.

tangential equation of Φ', which is the same as the equation of the reciprocal surface Φ in point coordinates a, b, c, d. We thus obtain the equation of Kummer's surface in the form of a symmetrical four-rowed determinant each element of which is linear in the coordinates. If the base points are taken as in the last paragraph the elements of the leading diagonal are zeros, and the equation has the form

$$\begin{vmatrix} 0 & z & y & x' \\ z & 0 & x & y' \\ y & x & 0 & z' \\ x' & y' & z' & 0 \end{vmatrix} = 0,$$

which is the same as

$$\sqrt{xx'} + \sqrt{yy'} + \sqrt{zz'} = 0,$$

where the letters represent linear functions of the coordinates a, b, c, d. In fact

$$x = aP_{23} + bQ_{23} + cR_{23} + dS_{23}$$

$$x' = aP_{14} + bQ_{14} + cR_{14} + dS_{14}$$

etc. Making use of the fact that the points (e) and (f) are base points we find that x, y, z, x', y', z' are connected by the two relations

$$e_2e_3x + e_3e_1y + e_1e_2z + e_1e_4x' + e_2e_4y' + e_3e_4z' = 0,$$

$$f_2f_3x + f_3f_1y + f_1f_2z + f_1f_4x' + f_2f_4y' + f_3f_4z' = 0,$$

which are of the kind required to make the general 14-nodal surface have two additional nodes (p. 88).

Incidentally we notice that the Jacobian of four quadrics having four, five, or six common points is birationally transformed by means of these quadrics into a surface whose reciprocal is a quartic surface having fourteen, fifteen, or sixteen nodes respectively.

On solving the first three of the four equations

$$\frac{\partial}{\partial x_s}(aP + bQ + cR + dS) = 0$$

for $x_1 : x_2 : x_3 : x_4$ we find

$$\frac{x_1}{x\phi} = \frac{x_2}{y\chi} = \frac{x_3}{z\psi} = \frac{x_4}{2xyz},$$

where
$$\phi = xx' - yy' - zz'$$
$$\chi = -xx' + yy' - zz'$$
$$\psi = -xx' - yy' + zz'$$
and the equation of Kummer's surface is expressible in the equivalent forms
$$\phi^2 - 4yy'zz' \equiv \chi^2 - 4zz'xx' \equiv \psi^2 - 4xx'yy' = 0;$$
we observe that the denominators represent linearly independent cubic surfaces containing the three concurrent conics in the tropes $x = 0$, $y = 0$, $z = 0$; this agrees with the first method of transformation.

Again, on solving the same three equations for $a : b : c : d$ we find these coordinates proportional to cubic functions of x_1, x_2, x_3, x_4 vanishing on the three lines $x_4 = 0$, $x_1x_2x_3 = 0$, and on a cubic curve lying on Weddle's surface. Thus plane sections of Kummer's surface correspond to sextic curves on Weddle's surface.

Caspary* gives the equation in terms of the tetrahedra whose corners are the nodes A, B, C, D, E, F and any point P of the surface, namely
$$PABC \, . \, PAEF \, . \, PBFD \, . \, PCDE = PBCD \, . \, PCAE \, . \, PABF \, . \, PDEF.$$

Cayley† gives the equation
$$3(xp_3 + zp_1 - 2t)\,\partial F/\partial x + (2zp_2 - tp_1)\,\partial F/\partial y + (xp_5 - 2yp_4)\,\partial F/\partial z$$
$$+ 3(2xp_6 - yp_5 - tp_3)\,\partial F/\partial t = 0,$$
where
$$F = 6xyzt - 4xz^3 - 4y^3t + 3y^2z^2 - x^2t^2,$$
and the six nodes are given by
$$x = y/\theta = z/\theta^2 = t/\theta^3,$$
$$f(\theta) = \theta^6 - p_1\theta^5 + p_2\theta^4 - p_3\theta^3 + p_4\theta^2 - p_5\theta + p_6 = 0.$$

A parametric expression of the surface is
$$x : y : z : t = U + V : vU + uV : v^2U + u^2V : v^3U + u^3V,$$
where
$$U^2 = f(u), \quad V^2 = f(v). \qquad \text{(Richmond.)}$$

* *Darboux Bulletin*, xv, 308.
† *Coll. Papers*, vii, 179.

CHAPTER XVI.

THETA FUNCTIONS.

§ 99. UNIFORMISATION OF THE SURFACE.

It is shown in the second chapter (p. 19) how the coordinates of any point on Kummer's surface may be expressed in terms of two parameters, which we now call x and x', by functions which are algebraic, but not uniform, since they involve the radicals

$$\sqrt{(x-k_1)(x-k_2)(x-k_3)(x-k_4)(x-k_5)(x-k_6)},$$

and $$\sqrt{(x'-k_1)(x'-k_2)(x'-k_3)(x'-k_4)(x'-k_5)(x'-k_6)}.$$

In other words, the points of Φ are represented uniformly by pairs of points on the curve

$$y^2 = f(x) \equiv (x-k_1)(x-k_2)(x-k_3)(x-k_4)(x-k_5)(x-k_6),$$

that is, by four variables, x, y, x', y', connected by two relations. If, then, we can express these four as uniform functions of two parameters, the uniformisation of the surface will be effected.

This is done by means of the integrals $\int dx/y$ and $\int x dx/y$, which are *finite* when taken along any portion of the curve. With arbitrary lower limits x_0, x_0', we put

$$v_1 = \int_{x_0}^{x} dx/y + \int_{x_0'}^{x'} dx'/y',$$

$$v_2 = \int_{x_0}^{x} x dx/y + \int_{x_0'}^{x'} x' dx'/y',$$

and then *invert*, that is, solve these equations for x and x'. It can be proved that $x + x'$ and xx' are *uniform periodic* functions of v_1 and v_2. The latter property is obvious since the integrals are indeterminate to the extent of additive multiples of their values when taken round the loops of the curve.

The *theta functions*, in terms of which the solution of the inversion problem may be expressed, arise in the attempt to construct periodic functions of two arguments in the form of doubly infinite series of exponentials. In the next section they are shown to give a uniform parametric representation; at present let the result of inversion be

$$x = \phi(v_1, v_2), \quad x' = \phi'(v_1, v_2).$$

If v_1 and v_2 are connected by the relation

$$\phi'(v_1, v_2) = x_0',$$

the second integrals disappear and we have

$$v_1 = \int_{x_0}^{x} dx/y, \quad v_2 = \int_{x_0}^{x} x\,dx/y,$$

so that the elimination of x from these two equations leads to $\phi'(v_1, v_2) = x_0'$, an equation which is really independent of x_0'. Thus the coordinates of any point on the sextic $y^2 = f(x)$ can be expressed in the form

$$x = \phi(v_1, v_2),$$

$$y = \frac{\partial\phi}{\partial v_1} - \frac{\partial\phi}{\partial v_2}\frac{\partial\phi'}{\partial v_1} \bigg/ \frac{\partial\phi'}{\partial v_2},$$

v_1 and v_2 being connected by

$$\phi'(v_1, v_2) = x_0'.$$

Now any quartic curve with one node can be transformed birationally into a sextic of this kind, and the k_s are projectively related to the tangents from the node. Since the six bitangents through any point of the singular surface of a quadratic complex are projectively related to the coefficients in the canonical equation of the complex, it follows that all the tangent sections of Kummer's surface can be transformed birationally into the *same* sextic and therefore into one another, and can be uniformly represented in terms of the same pair of integrals.

The chief use of this representation of plane curves lies in the application of a particular case of Abel's theorem. Namely of the result that the sum of integrals

$$\Sigma \int_{x_0}^{x} (ax + b)\, dx/y$$

has a constant value when the summation is extended to all the intersections of the sextic

$$y^2 = f(x) \equiv (x - k_1)(x - k_2)(x - k_3)(x - k_4)(x - k_5)(x - k_6)$$

with a variable algebraic curve of given order. The same theorem is true for any curve into which the sextic can be transformed birationally. The proof is elementary and may be given here. In the equation of the variable curve substitute $f(x)$ for y^2 so as to reduce it to the form

$$\phi(x) = y\psi(x),$$

and let the symbol δ refer to a change in its coefficients. The intersections are given by

$$F(x) \equiv \phi^2 - f\psi^2 = 0,$$

whence, on slightly varying the curve, the corresponding change in each intersection is given by

$$F'(x)\,\delta x + 2\phi\delta\phi - y^2 2\psi\delta\psi = 0,$$

which is the same as

$$\delta x/y = 2(\phi\delta\psi - \psi\delta\phi)/F'(x).$$

Now $\Sigma(ax+b)(\phi\delta\psi - \psi\delta\phi)/F'(x)$ vanishes when the summation is extended over all the roots of $F(x)=0$ because the degree in x of the numerator is at least two less than the degree of $F(x)$. Hence

$$\Sigma(ax+b)\,\delta x/y = 0,$$

which proves the theorem.

§ 100. DEFINITION OF THETA FUNCTIONS.

The functions which uniformise Kummer's surface are known as *theta functions**. We shall define them by their explicit expressions, and deduce their chief properties from these alone.

One of the greatest advances made by Jacobi in the theory of elliptic functions was the introduction of the singly infinite series $\Sigma \exp(an^2 + 2nu)$ as a uniform entire function possessing certain periodic properties. It is convenient to modify this slightly and write

$$\theta(u) = \Sigma \exp(2\pi i n u + \pi i \tau n^2),$$

the summation extending over all positive and negative integer values of n, and τ being any complex constant whose imaginary part is positive, to ensure convergence. There are three other functions, $\theta_{\alpha\beta}$, connected with this one and obtained from it by

* For information concerning these functions beyond what is required for the present purpose, and for references to the original authorities, see Baker, *Abelian Functions* (1896); Krazer, *Lehrbuch der Thetafunktionen* (1903).

replacing n by $n+\frac{1}{2}\alpha$ and u by $u+\frac{1}{2}\beta$, where α and β may be either 0 or 1; the ratios of these 'single' theta functions are elliptic functions.

We can generalise $\theta(u)$ without altering its formal expression, by interpreting the letters differently. n and u are now to be row-letters (see p. 25) and τ a symmetric square matrix, and we then have a multiply infinite series and a function of several arguments. In the case of a *double* theta function the general exponent, written in full, is

$$2\pi i(n_1u_1+n_2u_2)+\pi i(\tau_{11}n_1^2+2\tau_{12}n_1n_2+\tau_{22}n_2^2),$$

and the summation is extended over all integer values of n_1 and n_2. The condition for convergence is that the coefficient of i in $\tau_{11}n_1^2+2\tau_{12}n_1n_2+\tau_{22}n_2^2$ must be positive and not vanish for any values of n_1 and n_2 other than $n_1=n_2=0$.

With this $\theta(u)$ are associated fifteen other functions $\theta_{\alpha\beta}(u)$, obtained from $\theta(u)$ by replacing n_1, n_2, u_1, u_2 by $n_1+\frac{1}{2}\alpha_1$, $n_2+\frac{1}{2}\alpha_2$, $u_1+\frac{1}{2}\beta_1$, $u_2+\frac{1}{2}\beta_2$ respectively, where α_1, α_2, β_1, β_2 are integers. It is evident that these sixteen theta functions, being functions of only two arguments, must be connected by a great many relations. We proceed to find all these relations by elementary algebra and to coordinate them systematically by bringing them into connection with the orthogonal matrix considered in § 16.

§ 101. CHARACTERISTICS AND PERIODS.

By definition

$$\theta_{\alpha\beta}(u)=\Sigma\exp\{2\pi i(n+\tfrac{1}{2}\alpha)(u+\tfrac{1}{2}\beta)+\pi i\tau(n+\tfrac{1}{2}\alpha)^2\},$$

in which τ is a two-rowed symmetrical matrix,

$$\begin{pmatrix}\tau_{11} & \tau_{12}\\ \tau_{21} & \tau_{22}\end{pmatrix}\qquad(\tau_{12}=\tau_{21})$$

and all the other letters are row-letters, standing for pairs of letters distinguished by suffixes 1 and 2. The summation is for all integer values of n_1 and n_2, from $-\infty$ to $+\infty$. α_1 and α_2 are integers which may be taken to be either 0 or 1, since the integer parts of $\frac{1}{2}\alpha$ may be absorbed in n; β_1 and β_2 are also integers, and since the addition of even integers to β can at most change the sign of the function, it will generally be supposed that β_1 and β_2 are either 0 or 1. This being so, the matrix

$$\begin{pmatrix}\alpha_1 & \alpha_2\\ \beta_1 & \beta_2\end{pmatrix}$$

is called the *characteristic* of the theta function, and will be indicated by the suffix $\alpha\beta$. In accordance with the usual matrix notation $\alpha\beta$ denotes $\alpha_1\beta_1+\alpha_2\beta_2$, and the parity of the function depends on the value of this expression; for by taking a new pair of summation letters n', $=-n-\alpha$, we change the order of the terms without altering the value of the function, since the series is absolutely convergent; the general exponent is now

$$2\pi i(n'+\tfrac{1}{2}\alpha)(-u-\tfrac{1}{2}\beta)+\pi i\tau(n'+\tfrac{1}{2}\alpha)^2,$$

from which it follows that

$$\theta_{\alpha\beta}(-u)=(-)^{\alpha\beta}\,\theta_{\alpha\beta}(u).$$

A pair of quantities $\tau\bar{\alpha}+\bar{\beta}$, that is

$$\tau_{11}\bar{\alpha}_1+\tau_{12}\bar{\alpha}_2+\bar{\beta}_1,$$
$$\tau_{21}\bar{\alpha}_1+\tau_{22}\bar{\alpha}_2+\bar{\beta}_2,$$

is called a *period* on account of the periodic properties

$$\theta_{\alpha\beta}(u+\bar{\beta})=(-)^{\alpha\bar{\beta}}\,\theta_{\alpha\beta}(u),$$
$$\theta_{\alpha\beta}(u+\tau\bar{\alpha})=(-)^{\bar{\alpha}\beta}\exp(-2\pi i\bar{\alpha}u-\pi i\tau\bar{\alpha}^2)\,\theta_{\alpha\beta}(u).$$

The first of these is easily verifiable; the second depends on taking $n+\bar{\alpha}$ for a new pair of summation integers; thus the typical exponent on the left is

$$\begin{aligned}&2\pi i(n+\tfrac{1}{2}\alpha)(u+\tau\bar{\alpha}+\tfrac{1}{2}\beta)+\pi i\tau(n+\tfrac{1}{2}\alpha)^2\\&=2\pi i(n+\tfrac{1}{2}\alpha)(u+\tfrac{1}{2}\beta)+\pi i\tau(n+\tfrac{1}{2}\alpha+\bar{\alpha})^2-\pi i\tau\bar{\alpha}^2\\&=2\pi i(n'+\tfrac{1}{2}\alpha)(u+\tfrac{1}{2}\beta)+\pi i\tau(n'+\tfrac{1}{2}\alpha)^2-2\pi i\bar{\alpha}(u+\tfrac{1}{2}\beta)-\pi i\tau\bar{\alpha}^2,\end{aligned}$$

where $n'=n+\bar{\alpha}$, and this is the typical exponent on the right.

There are sixteen different characteristics, since each of the four elements α_1, α_2, β_1, β_2, may be 0 or 1. The one in which all the elements are 0 is called the zero characteristic. Corresponding to these there are sixteen *half periods* $\tfrac{1}{2}(\tau\alpha+\beta)$, and no two differ by a period. Any other half period differs by a period from one of these, and is said to be *congruent* to it. There are therefore only sixteen incongruent half periods. The effect of adding a half period to the argument of a theta function is to change the characteristic and multiply by a non-vanishing function. The formula

$$\theta_{\alpha\beta}(u+\tfrac{1}{2}\tau\bar{\alpha}+\tfrac{1}{2}\bar{\beta})=\exp\{-\pi i\bar{\alpha}(u+\tfrac{1}{2}\beta+\tfrac{1}{2}\bar{\beta}+\tfrac{1}{4}\tau\bar{\alpha})\}\,\theta_{\alpha+\bar{\alpha},\,\beta+\bar{\beta}}(u)$$

is easily verified by comparing the exponents for the same values of n, which are identical. We may say that the addition of half periods interchanges the thetas except as to exponential factors.

Of the sixteen characteristics six are odd and ten are even. Hence six thetas are odd functions and vanish for $u=0$. By adding the corresponding half periods to the arguments it follows from the last formula that the theta with zero characteristic vanishes for six half periods. Whence since the sum of two half periods is a half period, every theta vanishes for six half periods.

There is a close connection between the squares of the sixteen thetas and the sixteen linear forms considered in § 15. Accordingly we adopt a notation for the characteristics which brings out clearly the analogy. This is sufficiently indicated by the schemes

α_1, α_2	1, 0	1, 1	0, 1	0, 0
α	a	b	c	d

β_1, β_2	0, 1	1, 1	1, 0	0, 0
β	a	b	c	d

so that the six odd thetas are

$$\theta_{ab}, \theta_{ac}, \theta_{bc}, \theta_{ba}, \theta_{ca}, \theta_{cb}.$$

The *addition* of characteristics is effected by the addition of corresponding elements; after addition odd numbers are to be replaced by 1 and even numbers by 0. Hence, from either of the above schemes,

$$a+a=b+b=c+c=d+d=d,$$
$$b+c=a=a+d, \quad c+a=b=b+d, \quad a+b=c=c+d.$$

In other words, the symbols a, b, c, d, whether they stand for α or for β, obey the addition table

	a	b	c	d
a	d	c	b	a
b	c	d	a	b
c	b	a	d	c
d	a	b	c	d

which has a similar form to the multiplication table when a, b, c, d denoted a group of operations (p. 7). The sixteen characteristics can therefore be arranged in the form of an addition table:

dd	ac	ba	cb
ab	da	cc	bd
bc	cd	db	aa
ca	bb	ad	dc

It is evident that the half periods $\frac{1}{2}(\tau\alpha+\beta)$ obey the same laws of addition as the characteristics, and since θ_{dd} vanishes when $u=\frac{1}{2}(\tau a+b)$, etc., which are the half periods corresponding to the other symbols in the same column and row as dd, it follows that the same is true for any other θ, that is to say

$$\theta_{\alpha\beta}(\tfrac{1}{2}\tau\alpha'+\tfrac{1}{2}\beta')=0,$$

if $\alpha\beta$ and $\alpha'\beta'$ lie on the same row or on the same column. Hence what was before an incidence diagram becomes now a table of half period zeros of the theta functions.

§ 102. IDENTICAL RELATIONS AMONG THE DOUBLE THETA FUNCTIONS*.

The general exponent in the product $\theta_{\alpha\beta}(u)\,\theta_{\alpha\beta}(v)$ is

$$\begin{aligned}&2\pi i(m+\tfrac{1}{2}\alpha)(u+\tfrac{1}{2}\beta)+2\pi i(n+\tfrac{1}{2}\alpha)(v+\tfrac{1}{2}\beta)\\&\qquad+\pi i\tau(m+\tfrac{1}{2}\alpha)^2+\pi i\tau(n+\tfrac{1}{2}\alpha)^2\\&=\pi i(m+n+\alpha)(u+v+\beta)+\pi i(m-n)(u-v)\\&\qquad+\tfrac{1}{2}\pi i\tau(m+n+\alpha)^2+\tfrac{1}{2}\pi i\tau(m-n)^2.\end{aligned}$$

Now the pair of integers m, n can be of four different kinds as regards parity, for the remainders when divided by 2 may be $a, b, c,$ or d respectively. Write

$$\begin{aligned}m+n&=2\mu+\bar{\alpha},\\m-n&=2\nu\;+\bar{\alpha},\end{aligned}$$

where $\bar{\alpha}=a, b, c,$ or d. Then $\theta_{\alpha\beta}(u)\,\theta_{\alpha\beta}(v)$ may be arranged as the sum of four series in one of which the general exponent is

$$\begin{aligned}&2\pi i(\mu+\tfrac{1}{2}\alpha+\tfrac{1}{2}\bar{\alpha})(u+v+\beta)+2\pi i(\nu+\tfrac{1}{2}\bar{\alpha})(u-v)\\&\qquad+2\pi i\tau(\mu+\tfrac{1}{2}\alpha+\tfrac{1}{2}\bar{\alpha})^2+2\pi i\tau(\nu+\tfrac{1}{2}\bar{\alpha})^2.\end{aligned}$$

Since the summation is now with respect to the independent integers μ and ν, this leads to the product of two theta functions formed with periods 2τ instead of τ. If we write

$$\Theta_\alpha(u)=\Sigma\exp\{2\pi i(n+\tfrac{1}{2}\alpha)u+2\pi i\tau(n+\tfrac{1}{2}\alpha)^2\},$$

the terms in $\theta_{\alpha\beta}(u)\,\theta_{\alpha\beta}(v)$ for which $\bar{\alpha}$ is the same can be summed in the form

$$(-)^{\alpha\beta+\bar{\alpha}\beta}\,\Theta_{\alpha+\bar{\alpha}}(u+v)\,\Theta_{\bar{\alpha}}(u-v).$$

Hence giving $\bar{\alpha}$ the values a, b, c, d and adding the results, we find

$$\begin{aligned}\theta_{\alpha\beta}(u)\,\theta_{\alpha\beta}(v)&=\Sigma\,(-)^{\alpha\beta+\bar{\alpha}\beta}\,\Theta_{\alpha+\bar{\alpha}}(u+v)\,\Theta_{\bar{\alpha}}(u-v)\\&=\Sigma\,\Theta_{\alpha+\bar{\alpha}}(u-v)\,(-)^{\bar{\alpha}\beta}\,\Theta_{\bar{\alpha}}(u+v)\end{aligned}$$

* Cf. Clifford, "On the double theta-functions," *Coll. Papers*, p. 369; Baker, *Abelian Functions*, p. 526.

on rearranging the terms. By giving α and β the values a, b, c, d we obtain a matrix of sixteen elements which is evidently the product of the two matrices whose elements are

$$\Theta_{\alpha+\bar{\alpha}}(u-v) \qquad (\alpha, \bar{\alpha} = a, b, c, d),$$

and
$$(-)^{\bar{\alpha}\beta}\,\Theta_{\bar{\alpha}}(u+v) \qquad (\bar{\alpha}, \beta = a, b, c, d),$$

respectively. Written in full they are

$$\begin{bmatrix} \Theta_d(u-v), & \Theta_c(u-v), & \Theta_b(u-v), & \Theta_a(u-v) \\ \Theta_c(u-v), & \Theta_d(u-v), & \Theta_a(u-v), & \Theta_b(u-v) \\ \Theta_b(u-v), & \Theta_a(u-v), & \Theta_d(u-v), & \Theta_c(u-v) \\ \Theta_a(u-v), & \Theta_b(u-v), & \Theta_c(u-v), & \Theta_d(u-v) \end{bmatrix}$$

and
$$\begin{bmatrix} \Theta_a(u+v), & -\Theta_a(u+v), & -\Theta_a(u+v), & \Theta_a(u+v) \\ -\Theta_b(u+v), & \Theta_b(u+v), & -\Theta_b(u+v), & \Theta_b(u+v) \\ -\Theta_c(u+v), & -\Theta_c(u+v), & \Theta_c(u+v), & \Theta_c(u+v) \\ \Theta_d(u+v), & \Theta_d(u+v), & \Theta_d(u+v), & \Theta_d(u+v) \end{bmatrix}$$

and we recognise that they have the same form as the matrices which were multiplied together to give the sixteen linear forms (cf. pp. 29, 30). We infer that the sixteen products $\theta_{\alpha\beta}(u)\,\theta_{\alpha\beta}(v)$ are connected by exactly the same relations as the linear forms $(\alpha\beta)$. In particular they can be arranged as the elements of an orthogonal matrix

$$\begin{bmatrix} \theta_{dd}(u)\,\theta_{dd}(v), & \theta_{ac}(u)\,\theta_{ac}(v), & \theta_{ba}(u)\,\theta_{ba}(v), & \theta_{cb}(u)\,\theta_{cb}(v) \\ -\theta_{ab}(u)\,\theta_{ab}(v), & \theta_{da}(u)\,\theta_{da}(v), & -\theta_{cc}(u)\,\theta_{cc}(v), & \theta_{bd}(u)\,\theta_{bd}(v) \\ -\theta_{bc}(u)\,\theta_{bc}(v), & \theta_{cd}(u)\,\theta_{cd}(v), & \theta_{db}(u)\,\theta_{db}(v), & -\theta_{aa}(u)\,\theta_{aa}(v) \\ -\theta_{ca}(u)\,\theta_{ca}(v), & -\theta_{bb}(u)\,\theta_{bb}(v), & \theta_{ad}(u)\,\theta_{ad}(v), & \theta_{dc}(u)\,\theta_{dc}(v) \end{bmatrix}$$

From this, by giving v the values u and 0, nearly all the relations among the sixteen functions $\theta_{\alpha\beta}(u)$ can be deduced.

§ 103. PARAMETRIC EXPRESSION OF KUMMER'S SURFACE.

Firstly put $v = u$ and

$$x = \Theta_a(2u), \quad y = \Theta_b(2u), \quad z = \Theta_c(2u), \quad t = \Theta_d(2u),$$
$$x_0 = \Theta_a(0), \quad y_0 = \Theta_b(0), \quad z_0 = \Theta_c(0), \quad t_0 = \Theta_d(0),$$

then $\theta^2_{rs}(u)$ actually becomes the linear form denoted by (rs) in which $\alpha, \beta, \gamma, \delta$ have been replaced by x_0, y_0, z_0, t_0 respectively. Hence the squares of the sixteen theta functions satisfy all the identities which have been proved for the linear forms, and it is unnecessary here to enumerate them in detail. Any four $\theta^2_{rs}(u)$ which have a common half period for a zero are linearly connected, and all the quadratic relations which can be deduced from the linear forms are consequences of the statement that the sixteen $\theta^2_{rs}(u)$ can be arranged as the elements of an orthogonal matrix.

Secondly put $v=0$; this gives an orthogonal matrix

$$\begin{bmatrix} \theta_{dd}(0)\,\theta_{dd}(u), & 0, & 0, & 0 \\ 0, & \theta_{da}(0)\,\theta_{da}(u), & -\theta_{cc}(0)\,\theta_{cc}(u), & \theta_{bd}(0)\,\theta_{bd}(u) \\ 0, & \theta_{cd}(0)\,\theta_{cd}(u), & \theta_{db}(0)\,\theta_{db}(u), & -\theta_{aa}(0)\,\theta_{aa}(u) \\ 0, & -\theta_{bb}(0)\,\theta_{bb}(u), & \theta_{ad}(0)\,\theta_{ad}(u), & \theta_{dc}(0)\,\theta_{dc}(u) \end{bmatrix}$$

and the relations deduced from this show that all the irrational equations of Kummer's surface (§ 19) are identically satisfied after the preceding substitution. Hence

$$x=\Theta_a(2u),\quad y=\Theta_b(2u),\quad z=\Theta_c(2u),\quad t=\Theta_d(2u)$$

are the coordinates of any point on a Kummer surface expressed as uniform functions of two parameters u_1 *and* u_2.

The parameters of the nodes are deduced by a comparison with the algebraic representation. After substitution for the coordinates the equation of any trope becomes

$$\theta^2_{\alpha\beta}(u)=0,$$

so that we may say that $\theta_{\alpha\beta}(u)=0$ is the equation of one of the sixteen conics. Six of these equations are satisfied by each half period, and accordingly the half periods are the parameters of the nodes.

From their definitions, the functions to which the coordinates x, y, z, t are equated are theta functions of the arguments $2u$, constructed with constants 2τ instead of τ. The characteristics are ad, bd, cd, dd respectively, and are all even. Hence for all four functions

$$\Theta(-2u)=\Theta(2u),$$

and from the periodic property

$$\Theta(2u+2\tau\alpha+2\beta)=\exp(-2\pi i\alpha\,.\,2u-\pi i2\tau\alpha^2)\,\Theta(2u),$$

so that by the addition of a period $\tau\alpha+\beta$ to the arguments x, y, z, and t acquire the same exponential factor. Thus the ratios of the coordinates are quadruply periodic functions.

Every pair of values u_1, u_2 gives one point on the surface, but by what has just been proved, all the values

$$\pm u+\text{period}$$

give the same point. Additive periods will therefore be neglected, and then every point on the surface has *two* pairs of parameters $(\pm u)$ except the nodes which have only one.

The addition of a half period to (u) permutes the sixteen functions $\theta^2_{\alpha\beta}(u)$, save for exponential factors, in the same way as the group of operations of § 4 permutes the sixteen linear forms. Hence the collineations determined by the fundamental complexes are effected by adding the different half periods to the parameters.

§ 104. THETA FUNCTIONS OF HIGHER ORDER.

A *theta function of order* r and characteristic $(\alpha\beta)$ is defined as a one-valued entire analytic function satisfying the equation

$$\Theta^{(r)}_{\alpha\beta}(u+\tau\bar{\alpha}+\bar{\beta})=(-)^{\alpha\bar{\beta}+\bar{\alpha}\beta}\exp\{-2\pi i r\bar{\alpha}(u+\tfrac{1}{2}\tau\bar{\alpha})\}\,\Theta^{(r)}_{\alpha\beta}(u).$$

Obviously when the order and characteristic are given, the sum of any number of theta functions is another of the same kind. Again, from this equation it follows that the product of any two theta functions is another theta function, whose order and characteristic are obtained by adding those of its factors: thus

$$\Theta^{(r)}_{\alpha\beta}\,\Theta^{(s)}_{\alpha'\beta'}=\Theta^{(r+s)}_{\alpha+\alpha',\,\beta+\beta'}.$$

By repeating this process we find that the product of n of the sixteen theta functions of order 1 is a theta function of order n, whose characteristic is the sum of the n characteristics.

In consequence of the parametric expression of the surface, the terms of the irrational equation of any curve upon the surface become products of theta functions of the first order. The number of factors in each term is the order, n, of the equation, and the sum of the characteristics in any product is denoted by the symbol $(\alpha\beta)$, which was called the characteristic of the equation.

Further an odd (or even) product of radicals contains, after substitution, an odd (or even) number of odd thetas, and is therefore an odd (or even) function of u; so that the former qualifications of order, characteristic, and parity can now be taken to refer to theta functions. It follows from the properties of the irrational equation that

Every algebraic curve of order $2n$ on the general Kummer surface can be represented by an equation of the form

$$\Theta^{(n)}_{\alpha\beta}=0,$$

where Θ is a theta function of order n and characteristic $(\alpha\beta)$ and either odd or even.

The converse of this theorem is also true, namely that *every even or odd theta function, when equated to zero, represents an algebraic curve on the surface.* To prove this it is sufficient to show* that the number of linearly independent $\Theta_{\alpha\beta}^{(r)}$ is equal to the number of linearly independent curves belonging to the two families of order r and characteristic $(\alpha\beta)$, namely r^2.

From the defining equation it follows that

$$\Theta_{\alpha\beta}^{(r)}(u+\bar{\beta}) = (-)^{\alpha\bar{\beta}}\,\Theta_{\alpha\beta}^{(r)}(u),$$

so that $\Theta_{\alpha\beta}^{(r)}(u)$ is simply periodic in u_1 and in u_2 independently, the period being 2. Hence, by an extension of Fourier's theorem, it is possible to expand in the form

$$\Theta_{\alpha\beta}^{(r)}(u) \equiv \Sigma\Sigma A_{n_1 n_2}\, e^{\pi i (n_1 u_1 + n_2 u_2)} = \Sigma A_n\, e^{\pi i n u}, \text{ say.}$$

On substituting in the preceding condition and equating coefficients of $e^{\pi i (n_1 u_1 + n_2 u_2)}$, we find

$$A_{n_1 n_2}\, e^{\pi i n_1} = A_{n_1 n_2}\, e^{\pi i \alpha_1},$$

$$A_{n_1 n_2}\, e^{\pi i n_2} = A_{n_1 n_2}\, e^{\pi i \alpha_2},$$

so that only those terms occur for which $n_1 - \alpha_1$ and $n_2 - \alpha_2$ are both even. Further, on substituting in the condition,

$$\Theta_{\alpha\beta}^{(r)}(u+\tau\bar{\alpha}) = (-)^{\bar{\alpha}\beta} \exp\{-2\pi i r\bar{\alpha}(u + \tfrac{1}{2}\tau\bar{\alpha})\}\,\Theta_{\alpha\beta}^{(r)}(u),$$

and equating coefficients of $e^{\pi i n u}$ we find

$$e^{\pi i n \tau \bar{\alpha}} A_n = e^{\pi i \bar{\alpha}\beta - \pi i r \tau \bar{\alpha}^2} A_{n+2r\bar{\alpha}},$$

expressing $A_{n_1+2r,\, n_2}$ and $A_{n_1,\, n_2+2r}$ in terms of $A_{n_1 n_2}$. Hence $A_{n_1+2s,\, n_2+2t}$ may be chosen arbitrarily for $s, t = 0 \ldots r-1$ and then the remaining coefficients are determined. This proves that there are not more than r^2 linearly independent $\Theta_{\alpha\beta}^{(r)}$. Since however the equation of every curve leads to a theta function, the number must be exactly r^2.

From this point the properties of theta functions of any order may be deduced from the properties of the corresponding families of curves. For example the number of even functions is $\frac{1}{2}r^2$, $\frac{1}{2}(r^2 \pm 1)$, or $\frac{1}{2}(r^2 \pm 4)$ according to the nature of the order and characteristic. Another example is Poincaré's theorem† that the equations for u_1, u_2

$$\Theta_{\alpha\beta}^{(r)}(u+v) = 0, \quad \Theta_{\alpha'\beta'}^{(s)}(u+v') = 0$$

have $2rs$ common solutions. Consider first the case when $v = v' = 0$ and the functions have definite parity. Then the equations

* Baker, *Abelian Functions*, p. 452. † Reference, footnote p. 186 below.

represent curves of orders $2r$ and $2s$ intersecting at $2p$ nodes and $rs-p$ other points. Each of the latter gives two values of the arguments and each of the nodes only one, additive periods being always neglected. Hence the total number of solutions is

$$2(rs-p)+2p=2rs.$$

Again it follows at once from the definition that

$$\Theta^{(r)}_{\alpha\beta}(u+v)\,\Theta^{(r)}_{\alpha\beta}(u-v)\equiv\Theta^{(2r)}_{0}(u),$$

a theta function of order $2r$, zero characteristic and of definite parity if the functions on the left have also definite parity. Under a similar hypothesis

$$\Theta^{(s)}_{\alpha'\beta'}(u+v')\,\Theta^{(s)}_{\alpha'\beta'}(u-v')\equiv\Theta^{(2s)}_{0}(u),$$

and the $8rs$ common solutions of

$$\Theta^{(2r)}_{0}(u)=0,\quad \Theta^{(2s)}_{0}(u)=0$$

must be evenly divided among the pairs of equations

$$\left.\begin{matrix}\Theta^{(r)}_{\alpha\beta}(u+v)=0\\ \Theta^{(s)}_{\alpha'\beta'}(u+v')=0\end{matrix}\right\},\quad \left.\begin{matrix}\Theta^{(r)}_{\alpha\beta}(u+v)=0\\ \Theta^{(s)}_{\alpha'\beta'}(u-v')=0\end{matrix}\right\},$$

$$\left.\begin{matrix}\Theta^{(r)}_{\alpha\beta}(u-v)=0\\ \Theta^{(s)}_{\alpha'\beta'}(u+v')=0\end{matrix}\right\},\quad \left.\begin{matrix}\Theta^{(r)}_{\alpha\beta}(u-v)=0\\ \Theta^{(s)}_{\alpha'\beta'}(u-v')=0\end{matrix}\right\}.$$

§ 105. SKETCH OF THE TRANSCENDENTAL THEORY.

The whole subject may be approached from an entirely different point of view* by defining a *hyperelliptic surface* as one for which the coordinates of any point are proportional to uniform quadruply periodic functions of two parameters. It is then shown that the coordinates may be equated to *theta functions*†, and on the basis of certain fundamental propositions in transcendental analysis the geometrical theory of the surface is built up.

The hyperelliptic equation of an algebraic curve C on Kummer's surface Φ is obtained as follows. Let $S=0$ and $S'=0$ be two surfaces cutting Φ in the curve C, and in residual intersections having no common part. If the coordinates are replaced by theta functions, S/S' becomes a uniform quadruply periodic

* Humbert, "Théorie générale des surfaces hyperelliptiques," *Liouville*, sér. 4, IX, 29.

† Painlevé, *Comptes Rendus* (1902), CXXXIV, 808. See Krazer, *Lehrbuch der Thetafunktionen*, p. 126, and elsewhere for theorems and references.

function of the parameters u_1, u_2 and by a known theorem* can be expressed in the form ϕ/ϕ' where ϕ, ϕ' are uniform *entire* functions not simultaneously vanishing except where S/S' is indeterminate, which is not the case along C. Then by another theorem† $S/\phi = S'/\phi' = e^{g(u)} F(u)$, where F is a uniform entire function possessing the periodic properties

$$F(u_1+1,\ u_2) = F(u_1,\ u_2),$$
$$F(u_1,\ u_2+1) = e^{2\pi i a u} F(u),$$
$$F(u_1+\tau_{11},\ u_2+\tau_{21}) = e^{2\pi i (bu_1+cu_2+d)} F(u),$$
$$F(u_1+\tau_{12},\ u_2+\tau_{22}) = e^{2\pi i (b'u_1+c'u_2+d')} F(u).$$

By comparing the effects of adding periods in different orders we easily find that

$$a,\ b,\ b',\ c-a\tau_{11},\ c'-a\tau_{12},\ b\tau_{12}+c\tau_{22}-b'\tau_{11}-c'\tau_{21}$$

are integers. If we write $c = a\tau_{11}+e$, $c' = a\tau_{12}+e'$, the last result gives

$$-b'\tau_{11}+(b-e')\tau_{12}+e\tau_{22}+a(\tau_{11}\tau_{22}-\tau_{12}^2)+f=0,$$

where f is another integer. A relation of this form among the periods is called a *singular relation*; in general it is assumed that no such relation exists, in which case all the integer coefficients must vanish, and therefore $a=c=b'=0$, $b=c'$ and F becomes a theta function of order $-b$. Accordingly *the equation of every algebraic curve is obtained by equating a theta function to zero.* The converse proposition depends on the theorem that any three quadruply periodic functions are connected by an algebraic relation‡.

This is the theorem which was required in Chap. XIII. (p. 140) to fill the gap in a continuous algebraic theory of curves on the surface. A purely algebraical proof that the equation of *every* algebraic curve can be written in a certain irrational form when the constants k_s upon which the surface depends are perfectly general would probably be long and complicated, because the relations which may hold among the k_s for the theorem to fail are of many different forms, as will be seen in Chap. XVIII. On the other hand the transcendental expressions of the same relations have a perfectly definite form, only the integer coefficients in a linear relation being variable, so that an appeal to function-

* Poincaré, *Acta Math.* II, 97.

† Appell, *Liouville*, sér. 4, VII, 183, 196.

‡ Krazer, p. 116.

theory seems to be essential to the complete development of the subject.

One other theorem* is of fundamental importance, namely that two theta functions of orders m and n have $2mn$ common zeros, additive periods being disregarded. Then the surface obtained by equating the coordinates to theta functions, of the second order and zero characteristic, can be identified as a quartic surface with sixteen nodes, and other properties follow as already obtained in a more elementary manner. The tetrahedron of reference is here a fundamental tetrahedron.

By equating the coordinates to the squares of theta functions of the first order forming a Göpel tetrad and eliminating the arguments, an equation of the surface referred to a Göpel tetrad of tropes is obtained. This is the well-known *Göpel's biquadratic relation*† and corresponds to a rationalised form of

$$\sqrt{xx'} + \sqrt{yy'} + \sqrt{zz'} = 0,$$

after z and z' have been replaced by linear functions of

$$x,\ x',\ y,\ y'.$$

Again, if we take any one of the sixteen thetas of the first order, and equate the non-homogeneous coordinates to the negatives of

$$\partial^2 \log \theta/\partial u_2^2,\quad \partial^2 \log \theta/\partial u_2 \partial u_1,\quad \partial^2 \log \theta/\partial u_1^2$$

we obtain another parametric expression of Kummer's surface‡. If the fundamental sextic in k is

$$\lambda_0 + \lambda_1 k + \lambda_2 k^2 + \lambda_3 k^3 + \lambda_4 k^4 + \lambda_5 k^5 + \lambda_6 k^6,$$

by taking a new origin the equation of the surface can be written in the symmetrical determinant form

$$\begin{vmatrix} -\lambda_0 & \tfrac{1}{2}\lambda_1 & 2z & -2y \\ \tfrac{1}{2}\lambda_1 & -4z-\lambda_2 & 2y+\tfrac{1}{2}\lambda_3 & 2x \\ 2z & 2y+\tfrac{1}{2}\lambda_3 & -4x-\lambda_4 & \tfrac{1}{2}\lambda_5 \\ -2y & 2x & \tfrac{1}{2}\lambda_5 & -\lambda_6 \end{vmatrix} = 0.$$

The transcendental theory suggests two generalisations of Kummer's surface. By interpreting the matrix notation differently we can define theta functions of p variables $u_1 \ldots u_p$ with 2^{2p} different characteristics. These may be arranged in "Göpel

* Krazer, p. 42. Poincaré, *Bull. de la Soc. Math.*, XI, 129.

† Göpel, *Crelle* (1847), XXXV, 291. This is the historical origin of the transcendental theory. Baker, *Abelian Functions*, pp. 338, 466.

‡ Baker, *Proc. Camb. Phil. Soc.*, IX, 513 and XII, 219.

systems" of 2^p characteristics $(\alpha_s \beta_s)$. With one of these systems a p-fold in space of $2^p - 1$ dimensions is defined by

$$x_s = \theta^2_{\alpha_s \beta_s}(u), \quad (s = 1, 2, \dots 2^p),$$

a factor of proportionality being omitted. This p-fold possesses many properties analogous to those of Kummer's surface*.

A generalisation in a different direction† is suggested by the *hyperelliptic* theta functions, for which the elements of the symmetrical matrix τ are connected by $\frac{1}{2}(p-1)(p-2)$ relations, and the absolute constants upon which the function depends are the $2p-1$ independent cross ratios of the roots of the fundamental $(2p+2)$-ic equation. Selecting any one of the 2^{2p} thetas there are $\frac{1}{2}p(p+1)$ functions $\partial^2 \log \theta / \partial u_r \partial u_s$ $(r, s, = 1 \dots p)$ connected by $\frac{1}{2}p(p-1)$ relations. This set of relations represents a p-fold in space of $\frac{1}{2}p(p+1)$ dimensions.

* Wirtinger, (for $p=3$), *Göttinger Nachrichten* (1889), 474, and for the general case *Monatshefte für Mathematik und Physik* (1890), I, 113.

† Baker, *Proc. Camb. Phil. Soc.*, IX, 521. See also Klein, *Math. Ann.*, XXVIII, 557.

CHAPTER XVII.

APPLICATIONS OF ABEL'S THEOREM.

§ 106. TANGENT SECTIONS.

The hyperelliptic representation of the surface consists in equating the coordinates to theta functions of the second order and zero characteristic. The number of these which are linearly independent is 2^2, $= 4$, and so every equation of the form $\Theta_0^{(2)} = 0$ represents a plane section, as in the theory of quartic curves.

From the definition of theta functions of higher order it follows that

$$\Theta_{\alpha\beta}^{(r)}(u-v)\,\Theta_{\alpha\beta}^{(r)}(u+v) = \Theta_0^{2r}(u).$$

In particular, using theta functions of the first order,

$$\theta_{\alpha\beta}(u-v)\,\theta_{\alpha\beta}(u+v)$$

is a theta function of the second order and zero characteristic, with arguments u_1, u_2; hence the equation

$$\theta_{\alpha\beta}(u-v)\,\theta_{\alpha\beta}(u+v) = 0$$

represents a plane section. At every point of the section there are two pairs of parameters, (u_1, u_2) and $(-u_1, -u_2)$, one of which satisfies

$$\theta_{\alpha\beta}(u-v) = 0,$$

and the other

$$\theta_{\alpha\beta}(u+v) = 0,$$

so that either of these equations may be taken separately for the equation of the curve. The advantage of this is that by means of the equation

$$\theta_{\alpha\beta}(u-v) = 0$$

a *single* pair of parameters (u) is associated with each point of the curve, additive periods being neglected. Using x, y, z for non-homogeneous coordinates, in passing along the curve du_1/dx and du_2/dx are uniform quadruply periodic functions of u_1, u_2, and therefore expressible rationally in terms of x, y, z*; further, u_1 and u_2 are finite, so that $\int du_1$ and $\int du_2$ are "integrals of the first kind" for the curve. The number of such integrals is equal to the deficiency of the curve, and so the plane section is of deficiency 2 at least.

The equations

$$\theta(u-v)=0, \quad \theta(u+v)=0$$

have (p. 186) two common solutions, differing only in sign, and therefore representing the same point. Since either solution satisfies both of the equations

$$\frac{\partial}{\partial u_1}\{\theta(u-v)\,\theta(u+v)\}=0, \quad \frac{\partial}{\partial u_2}\{\theta(u-v)\,\theta(u+v)\}=0,$$

this point must be a double point on the curve. Hence *the section is by a tangent plane*, and has a double point at the point of contact.

In the equation of a tangent plane section it is indifferent which of the sixteen thetas is used. Selecting that with zero characteristic, we can easily find the coordinates of the plane. In the identity (p. 179)

$$\begin{aligned}\theta_0(u)\,\theta_0(v) &= \Theta_a(u-v)\,\Theta_a(u+v)+\Theta_b(u-v)\,\Theta_b(u+v)\\ &\quad +\Theta_c(u-v)\,\Theta_c(u+v)+\Theta_d(u-v)\,\Theta_d(u+v),\end{aligned}$$

replace u by $u+v$ and v by $u-v$; then

$$\begin{aligned}\theta_0(u+v)\,\theta_0(u-v) &= \Theta_a(2v)\,\Theta_a(2u)+\Theta_b(2v)\,\Theta_b(2u)\\ &\quad +\Theta_c(2v)\,\Theta_c(2u)+\Theta_d(2v)\,\Theta_d(2u)\\ &= \Theta_a(2v)\,x+\Theta_b(2v)\,y+\Theta_c(2v)\,z+\Theta_d(2v)\,t,\end{aligned}$$

so that the coordinates of the tangent plane are the same functions of v as the point coordinates of u, and $(\pm v)$ may be regarded as the parameters of the plane. This shows the self-polar nature of the surface with respect to the fundamental quadric $x^2+y^2+z^2+t^2=0$. Similarly by selecting another theta it may be shown that the surface is self-polar with respect to another quadric or a fundamental complex. Incidentally we notice that if $\theta(u\pm v)=0$, $(\theta=\text{odd})$, the line joining (u) and (v) is a ray of a fundamental complex, and the curve of intersection with a fundamental quadric is given by

$$\theta(2u)=0, \quad (\theta=\text{even}).$$

* Krazer, p. 117.

Let (u) be the point of contact of the plane (v), then

$$\theta(u-v)=0, \quad \theta(u+v)=0;$$

if $(P), =(P_1, P_2)$ is an odd period, the point $(v+\frac{1}{2}P)$ lies on the plane (v), and is the point of contact of the plane $(u+\frac{1}{2}P)$, which contains the point (u). Hence the line joining (u) and $(v+\frac{1}{2}P)$ is a *bitangent*, touching the surface at both points. Hence the six bitangents through the double point of the section

$$\theta_0(u-v)=0$$

touch the surface again at the points

$$u=v+\text{odd half period}.$$

Univocal curves.

The tangent sections are the simplest case of *univocal curves** on the surface, so named because of the two pairs of parameters of any point only *one* satisfies the equation of the curve. The general univocal curve is represented by

$$\Theta^{(n)}(u-v)=0,$$

where $\Theta(u)$ is a theta function of order n and zero characteristic, and (v) is any pair of constants. The other parameters of its points satisfy

$$\Theta^{(n)}(-u-v)=0,$$

and since $\Theta^{(n)}(u-v)\,\Theta^{(n)}(-u-v)$ is an even theta function of order $2n$ and zero characteristic, the univocal curve is of order $4n$ and the complete intersection with an n-ic surface. There are n^2 double points on the curve given by

$$\Theta^{(n)}(u-v)=0, \quad \Theta^{(n)}(-u-v)=0.$$

Thus the univocal curves occur among the ordinary curves on the surface, and are distinguished by the corresponding theta-function breaking into factors. If Θ and Θ' are of the same order and zero characteristic, consideration of the functions

$$\Theta(u-v)\,\Theta'(-u-v)\pm\Theta(-u-v)\,\Theta'(u-v)$$

leads without further analysis to numerous geometrical theorems of considerable interest.

§ 107. COLLINEAR POINTS.

We have seen that u_1 and u_2 are integrals of the first kind for the curve

$$\theta(u-v)=0,$$

* "Courbes univoques," Humbert, *Liouville*, 4, IX, 154.

and so, by Abel's theorem, the sum of the parameters of four collinear points is constant. If two of these points are at the double point, their parameters have zero sum; hence if (u) and (u') are collinear with the double point

$$u + u' = \text{const.}$$

Since u and u' are indeterminate to the extent of additive periods, this equation should be written

$$u + u' \equiv \text{const. (mod. } P).$$

By considering a bitangent we find that

$$u + u' \equiv 2v \text{ (mod. } P),$$

which gives another interpretation of the parameters of the tangent plane. This result may also be inferred from the fact that if (u) is any point on the curve, so also is (u'), $= (2v - u)$; for this establishes an involution on the plane quartic, and it is known that the only involution is that of points collinear with the double point. Bitangents of the surface are deduced from this by putting $u' = u$, giving

$$u = v + \tfrac{1}{2}P,$$

the half period being subject to the condition

$$\theta(\tfrac{1}{2}P) = 0.$$

The four points of intersection of the two tangent sections (v) and (v') are given by

$$\theta(u - v)\,\theta(u + v) = 0,$$
$$\theta(u - v')\,\theta(u + v') = 0,$$

and are the solutions of the four pairs of simultaneous equations

$$\left.\begin{array}{l}\theta(u - v) = 0\\ \theta(u - v') = 0\end{array}\right\}\ (1),\qquad \left.\begin{array}{l}\theta(u + v) = 0\\ \theta(u + v') = 0\end{array}\right\}\ (2),$$

$$\left.\begin{array}{l}\theta(u - v) = 0\\ \theta(u + v') = 0\end{array}\right\}\ (3),\qquad \left.\begin{array}{l}\theta(u + v) = 0\\ \theta(u - v') = 0\end{array}\right\}\ (4).$$

Let (a) and (b) be the two solutions of (1), $(-a)$ and $(-b)$ of (2), (c) and (d) of (3), $(-c)$ and $(-d)$ of (4). Then the four collinear points on $\theta(u - v) = 0$ have parameters (a), (b), (c), (d), and the same points on $\theta(u - v') = 0$ have parameters (a), (b), $(-c)$, $(-d)$. Hence any two tangent sections cut in four points, whose parameters on the two curves are the same except for two changes of sign.

If then (α), (β), (γ), (δ) are the (pairs of) parameters of four tangent planes through a line which cuts the surface in $(\pm a)$,

$(\pm b)$, $(\pm c)$, $(\pm d)$ we may suppose that the four points have parameters

$$\begin{array}{llllll} (a) & (-b) & (-c) & (d) & \text{on the section} & \theta(u-\alpha)=0, \\ (-a) & (b) & (-c) & (d) & \quad ,, \quad ,, & \theta(u-\beta)=0, \\ (-a) & (-b) & (c) & (d) & \quad ,, \quad ,, & \theta(u-\gamma)=0, \\ (a) & (b) & (c) & (d) & \quad ,, \quad ,, & \theta(u-\delta)=0. \end{array}$$

Since the sum of the parameters of four collinear points on $\theta(u-v)=0$ is $2v$ we have

$$\begin{aligned} a-b-c+d &= 2\alpha+P_1, \\ -a+b-c+d &= 2\beta+P_2, \\ -a-b+c+d &= 2\gamma+P_3, \\ a+b+c+d &= 2\delta+P_4. \end{aligned}$$

Substitute for $\alpha, \beta, \gamma, \delta$ in the conditions of incidence

$$\theta(a-\alpha)=0, \quad \theta(b-\beta)=0, \quad \theta(c-\gamma)=0, \quad \theta(d-\delta)=0,$$

then $$\theta\{\tfrac{1}{2}(a+b+c-d)-\tfrac{1}{2}P_s\}=0, \qquad (s=1, 2, 3, 4)$$

implying that four theta functions vanish for the same point $\frac{1}{2}(a+b+c-d)$. But singular conics do not intersect except at nodes, and so we infer that

$$\tfrac{1}{2}P_1 \equiv \tfrac{1}{2}P_2 \equiv \tfrac{1}{2}P_3 \equiv \tfrac{1}{2}P_4 \text{ (mod. } P).$$

Hence the conditions of collinearity of four points $(\pm a)$, $(\pm b)$, $(\pm c)$, $(\pm d)$ are that

$$\tfrac{1}{2}(-a+b+c+d), \quad \tfrac{1}{2}(a-b+c+d),$$
$$\tfrac{1}{2}(a+b-c+d), \quad \tfrac{1}{2}(a+b+c-d)$$

must all be zeros of the same theta function, that is, must represent points lying on the same singular conic. Of course, in these conditions, the signs of a, b, c, d may be changed.

If $2Q=P_1+P_2+P_3+P_4$ we find

$$\begin{aligned} a &\equiv \tfrac{1}{2}(\ \ \alpha-\beta-\gamma+\delta)+\tfrac{1}{2}Q, \\ b &\equiv \tfrac{1}{2}(-\alpha+\beta-\gamma+\delta)+\tfrac{1}{2}Q, \quad \text{(mod. } P) \\ c &\equiv \tfrac{1}{2}(-\alpha-\beta+\gamma+\delta)+\tfrac{1}{2}Q, \\ d &\equiv \tfrac{1}{2}(\ \ \alpha+\beta+\gamma+\delta)+\tfrac{1}{2}Q, \end{aligned}$$

and $$\tfrac{1}{2}(-a+b+c+d)=\tfrac{1}{2}(-\alpha+\beta+\gamma+\delta)+\tfrac{1}{2}Q,$$

so that the conditions for collinear planes are of the same form.

By way of illustration we shall consider the intersections of various kinds of lines with the surface. In the following, θ stands for any one of the sixteen theta functions.

(1) Let a tangent line at $(\pm u)$ meet the surface again at $(\pm a)$ and $(\pm b)$. Put $c = u$, $d = u$ and the four arguments become

$$\pm \tfrac{1}{2}(a-b) + u, \quad \pm \tfrac{1}{2}(a+b);$$

if however we put $c = u$, $d = -u$ the arguments become

$$\pm \tfrac{1}{2}(a-b), \quad \tfrac{1}{2}(a+b) \pm u.$$

This is practically the same case, since the sign of b is undetermined, and we have the conditions

$$\theta\left(\frac{a-b}{2}\right) = 0, \quad \theta\left(\frac{a+b}{2} + u\right) = 0, \quad \theta\left(\frac{a+b}{2} - u\right) = 0$$

for a_1, a_2, b_1, b_2, leaving one arbitrary. If $a + b = 2v$, then

$$\theta(a-v) = 0, \quad \theta(b-v) = 0, \quad \theta(u-v) = 0, \quad \theta(u+v) = 0,$$

showing that (v) is the tangent plane at (u), and contains (a) and (b).

(2) For a bitangent touching at (u) and (v) there are two cases. First put

$$a = b = u, \quad c = d = v.$$

The conditions of collinearity are

$$\theta(u) = 0, \quad \theta(v) = 0,$$

showing that the line is a chord of a singular conic. Secondly put $a = b = u$, $c = -d = v$; then the conditions of collinearity are

$$\theta(0) = 0, \quad \theta(u-v) = 0, \quad \theta(u+v) = 0,$$

showing that $(u-v)$ and $(u+v)$ are zeros of the same odd theta function. This is the result already obtained.

(3) For an inflexional tangent put $\pm a = \pm b = \pm c = u$. It will be found that there is only one distinct case, so we take $a = b = c = u$, $d = v$. Then the conditions are

$$\theta\left(\frac{u+v}{2}\right) = 0, \quad \theta\left(\frac{3u-v}{2}\right) = 0,$$

showing that $\pm \tfrac{1}{2}(u-v)$ is the parameter α of the tangent plane at u. Thus $v = u \pm 2\alpha$, and these are easily seen to be the points where the tangent lines at the double point to the tangent section cut the curve again.

(4) For a four-point contact tangent there are two cases according as the parameters are taken to be u, u, u, u, or $u, u, u, -u$. In the former case the condition is

$$\theta(u) = 0,$$

and the line touches a conic. In the second case

$$\theta(0)=0 \text{ and } \theta(2u)=0;$$

whence the locus of the points of contact is one of the six curves

$$\theta_{\alpha\beta}(2u)=0,$$

where $(\alpha\beta)$ is an odd characteristic. These are therefore the equations of the principal asymptotic curves.

§ 108. ASYMPTOTIC CURVES.

Let (u) be the point of contact of the plane (v), then

$$\theta(u+v)=0, \quad \theta(u-v)=0,$$

and simultaneous increments are connected by

$$\theta^{(1)}(u+v)(du_1+dv_1)+\theta^{(2)}(u+v)(du_2+dv_2)=0,$$
$$\theta^{(1)}(u-v)(du_1-dv_1)+\theta^{(2)}(u-v)(du_2-dv_2)=0,$$

where $\theta^{(1)}$ and $\theta^{(2)}$ are the partial derivatives with respect to the first and second arguments.

The inflexional tangents at the point (u) are the directions of the two branches of the tangent section

$$\theta(u \pm v)=0.$$

One of these is given by

$$\theta^{(1)}(u+v)\,du_1+\theta^{(2)}(u+v)\,du_2=0.$$

Making use of this we find that either

$$\begin{vmatrix} \theta^{(1)}(u+v), & \theta^{(2)}(u+v) \\ \theta^{(1)}(u-v), & \theta^{(2)}(u-v) \end{vmatrix}=0,$$

or $\qquad du_1=dv_1$

and $\qquad du_2=dv_2$.

From the first alternative we deduce

$$\theta^{(1)}(u-v)\,du_1+\theta^{(2)}(u-v)\,du_2=0,$$

showing that the two inflexional tangents coincide and the locus of u is either a cusp locus or an envelope of asymptotic curves: the cusp locus consists of isolated points at the nodes, and the envelope, which is the parabolic curve, consists of the sixteen singular conics. If (u) is a general point of the surface we must take the second alternative and find the integrated equation*

$$u=v+k.$$

* Reichardt, *Nova Acta Leopoldina* (1886), L. Hutchinson, *Amer. Bull.* (1899), V, 465.

Hence the hyperelliptic equation of an asymptotic curve is

$$\theta(2u - k) = 0,$$

where the constant k satisfies

$$\theta(k) = 0.$$

The sixteen points $(k + \frac{1}{2}P)$ are the points of contact of the curve with the tropes.

The equation should be written

$$\theta(2u + k)\,\theta(2u - k) = 0,$$

and it is easily verified that this product is an even theta function of order 8 and zero characteristic, and therefore the asymptotic curve is an algebraic curve of order 16 and the complete intersection of a quartic surface passing through all the nodes. Thus the asymptotic curves do not belong to the "singular" family, for these are given by the vanishing of an *odd* theta function.

The asymptotic curve has sixteen double points, beside the cusps at the nodes, given by

$$\theta(2u + k) = 0,$$
$$\theta(2u - k) = 0;$$

these points are obtained from any one of them by the addition of half-periods to its parameters, that is by the group of sixteen collineations.

The equations

$$\theta(2u - k) = 0, \quad \theta(k) = 0,$$

regarded as equations for (k), have two solutions, the parameters of the two asymptotic curves through any point (u); if (k') is one, then $(k') = (2u - k)$ is the other. Hence the curves, whose parameters are (k) and (k'), cut in the points given by

$$2u = k + k',$$

that is, in the sixteen points $\frac{1}{2}(k + k') + \frac{1}{2}P$. Since the equation of the second curve may be taken in the form $\theta(2u + k') = 0$, it follows that the sixteen points $\frac{1}{2}(k - k') + \frac{1}{2}P$ are also on both curves. The points $\frac{1}{2}(k + k')$ and $\frac{1}{2}(k - k')$ are the points of contact of a bitangent, and so we have here the configuration of thirty-two points obtained from one point by drawing a succession of bitangents, and corresponding to the projections of any point on Weddle's surface from the nodes.

For six special values of (k), namely half-periods satisfying $\theta(k) = 0$, the two factors of $\theta(2u + k)\,\theta(2u - k)$ become essentially the same, and we get the six principal asymptotic curves of order 8. We have

$$\theta_{\alpha\beta}(u+\tau\bar{\alpha}+\bar{\beta})=(-)^{\alpha\bar{\beta}+\bar{\alpha}\beta}\exp(-2\pi i\bar{\alpha}u-\pi i\tau\bar{\alpha}^2)\,\theta_{\alpha\beta}(u),$$
$$\theta_{\alpha\beta}(2u+2\tau\bar{\alpha}+2\bar{\beta})=\exp(-8\pi i\bar{\alpha}u-4\pi i\tau\bar{\alpha}^2)\,\theta_{\alpha\beta}(2u),$$

and $\theta_{\alpha\beta}(2u)$ is a theta function of order 4 and zero characteristic, and of the same parity as $\theta_{\alpha\beta}(u)$. Hence the principal asymptotic curves belong to the family cut out by quartic surfaces through four conics.

As an example of the application of Abel's theorem to curves on the surface the following may be given. If a $4n$-ic curve passes through all the nodes, any tangent plane cuts it in $4n$ points, which together with the point of contact lie on an infinite number of plane $(n+1)$-ics; each of the latter cuts the tangent section again in two points collinear with the point of contact. If the point of contact is on the $4n$-ic the remaining $4n-2$ intersections lie on a plane n-ic through the point of contact.

§ 109. INSCRIBED CONFIGURATIONS.

The power of this method is well shown by the ease with which certain inscribed and circumscribed configurations may be constructed. Only a few examples are given here.

It must be remembered that in speaking of a point (u) of the surface we mean the point whose parameters are $(\pm u_1, \pm u_2)$ to which any (pair of) periods may be added.

We have seen that the points (a), (b), (c), (d) are collinear if the points (x), (y), (z), (t) lie on the same conic, where

$$\begin{aligned} 2x &= -a+b+c+d \\ 2y &= a-b+c+d \\ 2z &= a+b-c+d \\ 2t &= a+b+c-d, \end{aligned}$$

and that then the four tangent planes through the line are (α), (β), (γ), (δ), where

$$\begin{aligned} 2\alpha &= a-b-c+d \\ 2\beta &= -a+b-c+d \\ 2\gamma &= -a-b+c+d \\ 2\delta &= a+b+c+d; \end{aligned}$$

thus

$$\begin{aligned} x &= d-\alpha = c+\beta = b+\gamma = -a+\delta \\ y &= c+\alpha = d-\beta = a+\gamma = -b+\delta \\ z &= b+\alpha = a+\beta = d-\gamma = -c+\delta \\ t &= a-\alpha = b-\beta = c-\gamma = -d+\delta, \end{aligned}$$

showing all the incidences; for the condition that a plane (v) may contain a point (u) is that $(u \pm v)$ be a point on a certain conic.

Klein's tetrahedra.

If now we suppose that (t) does not lie on the conic, but is arbitrary, all the incidences still hold except those indicated by the last line, so that the planes and points are the faces and corners of a tetrahedron. These are Klein's "principal tetrahedra" (§ 34). x_1, y_1, z_1, t_1, t_2 are arbitrary and so the number of tetrahedra is ∞^5.

Instead of taking (x), (y), (z) arbitrarily on the conic $\theta(u) = 0$ we might have taken (a), (b), (c) arbitrarily on the section

$$\theta(u - \delta) = 0;$$

then the equations

$$c + \alpha = -b + \delta = y$$
$$b + \alpha = -c + \delta = z$$

show that the planes (α), (δ) are "conjugate" to the points (b), (c), for when α, δ, b are given c is uniquely determined and therefore the corresponding point is rationally determined.

From the relations between the points and planes, having regard to the fact that the signs are indeterminate, we see that there is not a one-one correspondence between a, b, c, d and $\alpha, \beta, \gamma, \delta$ but the three partitions into two pairs correspond, giving six conjugate sets, incident with the six edges of the principal tetrahedron.

Expressed in terms of x, y, z, t, the points and planes are

$$\begin{aligned} 2a &= -x + y + z + t & \qquad 2\alpha &= -x + y + z - t \\ 2b &= x - y + z + t & 2\beta &= x - y + z - t \\ 2c &= x + y - z + t & 2\gamma &= x + y - z - t \\ 2d &= x + y + z - t & 2\delta &= x + y + z + t \end{aligned}$$

and a typical conjugate set is given by

$$\left.\begin{aligned} \alpha &= -\delta + y + z \\ \delta &= \delta \end{aligned}\right\} \qquad \left.\begin{aligned} b &= \delta - y \\ c &= \delta - z \end{aligned}\right\}$$

where δ is arbitrary and $\theta(y) = 0 = \theta(z)$.

Rohn's theorem.

The condition of incidence of point (u) and plane (v) being

$$\theta(u \pm v) = 0$$

we can write down the parameters of sixteen points of Kummer's surface and sixteen tangent planes forming in themselves a 16_6

configuration. Let (a), (b), (c), (d), (e), (f) be any six points on the conic

$$\theta(u) = 0;$$

then the plane $\quad \tfrac{1}{2}(a+b+c+d+e+f), \; = v$

contains the six points

$$\tfrac{1}{2}(-a+b+c+d+e+f) \ldots\ldots\ldots \tfrac{1}{2}(a+b+c+d+e-f)$$

and these lie in the planes obtained from the first by changing the signs of two of the parameters, and so on. The group of operations is precisely the same as that which deduces thirty-two lines from a given one in Klein coordinates. The sum of the parameters of the above six coplanar points is $4v$, showing that they lie on a conic through the point of contact of their plane*.

Humbert's tetrahedra.

The parametric representation of Humbert's tetrahedra† (p. 158) is obtained in an equally simple manner by taking any three points (x), (y), (z) on the conic

$$\theta(u) = 0.$$

The plane $(y+z)$ contains the point (y) since the difference of the parameters represents a point on the conic, and for a similar reason contains (z). The second tangent plane through these two points is $(y-z)$. The planes $(x-y)$, $(y+z)$, $(z-x)$ meet in the point $(-x+y+z)$, and so on; thus the six tangent planes through the sides of the triangle (x), (y), (z) meet by threes in the points

$$\begin{aligned}(a) &= (-x+y+z)\\(b) &= (x-y+z)\\(c) &= (x+y-z)\\(d) &= (x+y+z)\end{aligned}$$

which are the corners of an inscribed tetrahedron. It remains to be shown that the edges are tangent lines. Now the tangent plane $(y+z)$ contains the points (a) and (d), and since

$$(a)+(d) = 2(y+z),$$

the line joining them passes through the point of contact. Similarly for the other edges.

* This is Rohn's 'first theorem,' see *Math. Ann.*, xv, 350. Generalisations and extensions are suggested by Klein, *Math. Ann.*, xxvii, 106, where the transcendental representation is deduced directly from line coordinates without the introduction of theta functions.

† Humbert, *Liouville*, sér. 4, ix, 123, where generalisations are given.

Since the configuration is determined by three arbitrary points on a conic, the number of such tetrahedra is triply infinite. It can be shown that no other inscribed tetrahedra have the property that their edges are tangent lines.

The pairs of parameters (u) of points on a conic $\theta(u) = 0$ may be taken to represent the points on a plane quartic curve with one node, in such a way that points collinear with the node differ only in the signs of their parameters. The points of contact of the tangents from the node are represented by the half-periods which are zeros of $\theta(u)$. We may take a cubic curve in space and establish a correspondence between its points and the pencil of lines through the node of the plane quartic*. A group of three points on the quartic determines three lines of the pencil and three points of the cubic and therefore the plane joining them. Consider the groups of three points cut out by conics through the node and three fixed points of the quartic; they depend on one variable and hence the corresponding planes in space form a developable. Since a given line through the node determines two groups of the series, therefore two planes of the developable pass through a given point, and the developable is a quadric cone. This cone meets the cubic at points for which the two planes of the developable coincide and which therefore correspond to the tangents to the quartic from the node. Thus the cone passes through six fixed points and the locus of the vertex is Weddle's surface.

The group of points (x) (y) (z) on the quartic belongs to a linear series if, by Abel's theorem, $(x+y+z)$ has a constant value. Since at the same time $(-x-y-z)$ has a constant value, it follows that the groups $(-x)$ $(-y)$ $(-z)$, projections of the former groups from the node, belong to another linear series. These two series determine the same point on Weddle's surface, which may be denoted by the pair of parameters $(x+y+z)$ or by $(-x-y-z)$. By comparison with the parametric representation of Humbert's tetrahedra we see that if AA', BB', CC' are the intersections of the quartic curve with any three lines through the node, the corners of a Humbert's tetrahedron are represented by the pairs of groups of points

$$\left.\begin{matrix} ABC \\ A'B'C' \end{matrix}\right\} \quad \left.\begin{matrix} A'BC \\ AB'C' \end{matrix}\right\} \quad \left.\begin{matrix} AB'C \\ A'BC' \end{matrix}\right\} \quad \left.\begin{matrix} ABC' \\ A'B'C \end{matrix}\right\}.$$

* Wirtinger, *Jahresbericht der Deutschen Mathematiker Vereinigung*, IV, 97.

CHAPTER XVIII.

SINGULAR KUMMER SURFACES.

§ 110. ELLIPTIC SURFACES.

We have seen in the case of the Wave surface that in consequence of a special situation of the nodes in each trope and a corresponding relation among the coefficients k_s of the quadratic complex, the point coordinates may be expressed in terms of elliptic functions. In the present section we seek the corresponding relation among the periods of the theta functions. Starting with the more general problem of linearly transforming the arguments of the theta functions into arguments of elliptic functions, we find that the relation among the periods has a certain definite form, and by examining the different relations of this form we are led to a series of *elliptic* Kummer surfaces of which the Wave surface is the first, and also to other *singular* Kummer surfaces of which elliptic surfaces are particular cases.

Consider the general theta function $\theta(u_1, u_2)$ of the first order for which a general pair of periods is

$$\begin{array}{c|l} u_1 & \tau_{11}\alpha_1 + \tau_{12}\alpha_2 + \beta_1 \\ u_2 & \tau_{21}\alpha_1 + \tau_{22}\alpha_2 + \beta_2 \end{array}$$

where $\alpha_1, \alpha_2, \beta_1, \beta_2$ are any integers. We seek the conditions that it may be possible to take linear combinations of u_1 and u_2 for new arguments U, V so that θ may be expressible in terms of elliptic functions of U and other elliptic functions of V; that is, so that θ may be doubly periodic in U alone and in V alone.

Assume $$U = g_1 u_1 + g_2 u_2,$$

then the four periods of U must be linear combinations of the two periods Ω and Ω' of the elliptic functions of U. Hence eight *integers* m_s, m_s' can be found so that

$$\begin{aligned} g_1\tau_{11} + g_2\tau_{21} &= m_1\Omega + m_1'\Omega' \\ g_1\tau_{12} + g_2\tau_{22} &= m_2\Omega + m_2'\Omega' \\ g_1 \qquad\quad &= m_3\Omega + m_3'\Omega' \\ g_2 &= m_4\Omega + m_4'\Omega'; \end{aligned}$$

whence, on eliminating g_1, g_2, Ω, Ω',

$$\begin{vmatrix} \tau_{11} & \tau_{21} & m_1 & m_1' \\ \tau_{12} & \tau_{22} & m_2 & m_2' \\ 1 & 0 & m_3 & m_3' \\ 0 & 1 & m_4 & m_4' \end{vmatrix} = 0,$$

or, writing $p_{rs} = m_r m_s' - m_s m_r'$,

$$p_{23}\tau_{11} + (p_{31} + p_{24})\tau_{12} + p_{41}\tau_{22} + p_{43}(\tau_{12}^2 - \tau_{11}\tau_{22}) + p_{12} = 0.$$

Hence a necessary condition is that the quantities

$$\tau_{11},\ \tau_{12},\ \tau_{22},\ \tau_{12}^2 - \tau_{11}\tau_{22},$$

be connected by a linear relation with integer coefficients

$$A\tau_{11} + B\tau_{12} + C\tau_{22} + D(\tau_{12}^2 - \tau_{11}\tau_{22}) + E = 0,$$

and from the identity $\qquad p_{14}p_{23} + p_{24}p_{31} + p_{34}p_{12} = 0$

we deduce $\qquad B^2 - 4AC - 4DE = (p_{31} - p_{24})^2, \ = k^2$, say,

where k is an integer. Conversely, when $B^2 - 4AC - 4DE$ is the square of a given integer k, the p_{rs} are uniquely determined in terms of the coefficients in the linear relation, and from them the integers m_s, m_s', with a certain degree of arbitrariness.

§ 111. TRANSFORMATION OF THETA FUNCTIONS.

If we replace the four pairs of periods (τ_{11}, τ_{21}), (τ_{12}, τ_{22}), $(1, 0)$, $(0, 1)$ by any four independent linear combinations, and take new arguments, linear functions of u_1 and u_2, so that the new periods are *normal*, that is, have the form (τ_{11}', τ_{21}'), (τ_{12}', τ_{22}'), $(1, 0)$, $(0, 1)$; then these new periods can be used to construct a new theta function, provided $\tau_{12}' = \tau_{21}'$.

Take any sixteen integers, elements of four two-rowed matrices α, β, γ, δ, and form with them four pairs of periods

$$\begin{array}{c|l} u_1 & \tau_{11}\alpha_{11} + \tau_{12}\alpha_{21} + \beta_{11},\quad \tau_{11}\alpha_{12} + \tau_{12}\alpha_{22} + \beta_{12}, \\ & \tau_{11}\gamma_{11} + \tau_{12}\gamma_{21} + \delta_{11},\quad \tau_{11}\gamma_{12} + \tau_{12}\gamma_{22} + \delta_{12}, \\ u_2 & \tau_{21}\alpha_{11} + \tau_{22}\alpha_{21} + \beta_{21},\quad \tau_{21}\alpha_{12} + \tau_{22}\alpha_{22} + \beta_{22}, \\ & \tau_{21}\gamma_{11} + \tau_{22}\gamma_{21} + \delta_{21},\quad \tau_{21}\gamma_{12} + \tau_{22}\gamma_{22} + \delta_{22}, \end{array}$$

or, in matrix notation, $\qquad u \mid \tau\alpha + \beta,\ \ \tau\gamma + \delta.$

Next take new arguments

$$v = (\tau\gamma + \delta)^{-1} u$$

so as to reduce the last two pairs of periods to the normal form

$$\begin{pmatrix} 1 & 0 \\ 0 & 1 \end{pmatrix}.$$

Then the first two pairs of periods are given by the columns of the matrix

$$\tau' = (\tau\gamma + \delta)^{-1} (\tau\alpha + \beta).$$

The condition for a *transformation* is that τ' must be a symmetrical matrix; on equating it to its conjugate we have

$$(\tau\gamma + \delta)^{-1} (\tau\alpha + \beta) = (\bar{\alpha}\tau + \bar{\beta}) (\bar{\gamma}\tau + \bar{\delta})^{-1},$$

whence $$(\tau\alpha + \beta)(\bar{\gamma}\tau + \bar{\delta}) = (\tau\gamma + \delta)(\bar{\alpha}\tau + \bar{\beta}),$$

or $$\tau(\alpha\bar{\gamma} - \gamma\bar{\alpha})\tau + (\beta\bar{\gamma} - \delta\bar{\alpha})\tau + \tau(\alpha\bar{\delta} - \gamma\bar{\beta}) + \beta\bar{\delta} - \delta\bar{\beta} = 0,$$

which leads to a single relation of the form

$$A\tau_{11} + B\tau_{12} + C\tau_{22} + D(\tau_{12}^2 - \tau_{11}\tau_{22}) + E = 0,$$

the coefficients being integers. Now in the case of an *ordinary* transformation* it is assumed that no relation of this form exists, so that we must have

$$\alpha\bar{\gamma} - \gamma\bar{\alpha} = 0$$
$$\beta\bar{\delta} - \delta\bar{\beta} = 0$$
$$\delta\bar{\alpha} - \beta\bar{\gamma} = r = \alpha\bar{\delta} - \gamma\bar{\beta},$$

where by r is to be understood a numerical multiple of the unit matrix. These represent five conditions for the sixteen integers. The integer r is called the *order* of the transformation.

These equations can be written in another form which is often useful. As they stand they express that

$$\begin{pmatrix} \delta & \beta \\ \gamma & \alpha \end{pmatrix} \begin{pmatrix} \bar{\alpha} & -\bar{\beta} \\ -\bar{\gamma} & \bar{\delta} \end{pmatrix} = \begin{pmatrix} r & 0 \\ 0 & r \end{pmatrix},$$

each matrix having four rows; since the right side is a numerical multiple of the unit matrix, the order of the factors on the left may be reversed and we have

$$\begin{pmatrix} \bar{\alpha} & -\bar{\beta} \\ -\bar{\gamma} & \bar{\delta} \end{pmatrix} \begin{pmatrix} \delta & \beta \\ \gamma & \alpha \end{pmatrix} = \begin{pmatrix} r & 0 \\ 0 & r \end{pmatrix},$$

which is equivalent to the equations

$$\bar{\alpha}\delta - \bar{\beta}\gamma = r = \bar{\delta}\alpha - \bar{\gamma}\beta,$$
$$\bar{\alpha}\beta - \bar{\beta}\alpha = 0 = \bar{\gamma}\delta - \bar{\delta}\gamma.$$

* As considered by Hermite, *Comptes Rendus* (1855), XL, 249. A reproduction of this memoir is given by Cayley, *Quarterly Journal*, XXI, 142 ; *Coll. Papers*, XII, 358.

§ 112. THE INVARIANT.

When a relation of the form

$$A\tau_{11} + B\tau_{12} + C\tau_{22} + D(\tau_{12}^2 - \tau_{11}\tau_{22}) + E = 0$$

exists, the corresponding theta functions are called *singular*. The Kummer surface represented by them is also called singular, and possesses geometrical features which are absent in the general case. It is evident from what precedes that a singular theta-function may allow a transformation which does not exist when the periods are arbitrary; with such *singular transformations* we are not here concerned*.

We have next to prove that the preceding relation is changed, by any transformation, into another of the same form, and that $\Delta, = B^2 - 4AC - 4DE$, *is an invariant for linear transformations.* Consider first the elliptic case. Let the new periods of $g_1u_1 + g_2u_2$ be written in the form $M_s\Omega + M_s'\Omega'$, then

$$M_1\Omega + M_1'\Omega' = (m_1\Omega + m_1'\Omega')\,\alpha_{11} + (m_2\Omega + m_2'\Omega')\,\alpha_{21} + (m_3\Omega + m_3'\Omega')\,\beta_{11} + (m_4\Omega + m_4'\Omega')\,\beta_{21},$$

and so on; thus $M_1 \ldots M_4'$ are integers given by

$$\begin{pmatrix} M_1 & M_2 & M_3 & M_4 \\ M_1' & M_2' & M_3' & M_4' \end{pmatrix} = \begin{pmatrix} m_1 & m_2 & m_3 & m_4 \\ m_1' & m_2' & m_3' & m_4' \end{pmatrix} \begin{pmatrix} \alpha & \gamma \\ \beta & \delta \end{pmatrix}.$$

From this and from the relations satisfied by $\alpha, \beta, \gamma, \delta$ we deduce

$$\begin{vmatrix} M_1 & M_3 \\ M_1' & M_3' \end{vmatrix} - \begin{vmatrix} M_2 & M_4 \\ M_2' & M_4' \end{vmatrix} = (p_{13} - p_{24})\,r.$$

Now the new periods are connected by the relation

$$\begin{vmatrix} \tau_{11}' & \tau_{21}' & M_1 & M_1' \\ \tau_{12}' & \tau_{22}' & M_2 & M_2' \\ 1 & 0 & M_3 & M_3' \\ 0 & 1 & M_4 & M_4' \end{vmatrix} = 0,$$

which may be expanded in the form

$$\rho\,[A'\tau_{11}' + B'\tau_{12}' + C'\tau_{22}' + D'(\tau_{12}'^2 - \tau_{11}'\tau_{22}') + E'] = 0,$$

making allowance for a possible extraneous integer factor ρ. This is a singular relation of the same form as before, and the new invariant is

$$\Delta' = B'^2 - 4A'C' - 4D'E',$$

whence

$$\rho^2\Delta' = (p_{13} - p_{24})^2 r^2 = r^2\Delta.$$

* See Humbert, "Sur les fonctions abéliennes singulières (deuxième mémoire)," *Liouville*, sér. 5, VI, 279.

If $r=1$, the inverse transformation is also linear, so that we have in this case $\sigma^2\Delta=\Delta'$, where $\sigma \gtrless 1$; whence $\rho=1=\sigma$ and $\Delta'=\Delta$, proving the invariance.

In the general singular case in which Δ is not a square, m_s, m_s', not necessarily integers, can be found in many ways to express the given singular relation in determinant form. The lines of the preceding proof may then be followed and lead to the same two properties of formal and numerical invariance.

It can be proved* that by a linear transformation the singular relation can be reduced to one of the canonical forms

$$-\tfrac{1}{4}\Delta\tau_{11}+\tau_{22}=0,$$
$$-\tfrac{1}{4}(\Delta-1)\tau_{11}+\tau_{12}+\tau_{22}=0,$$

according to the parity of B. Both of these are included in the form

$$A\tau_{11}+B\tau_{12}+C\tau_{22}=0,$$

from which it follows that *the invariant is always positive*; for

$$\begin{aligned}\Delta\tau_{22}&=(B^2-4AC)\tau_{22}+4A(A\tau_{11}+B\tau_{12}+C\tau_{22})\\&=4A^2\tau_{11}+4AB\tau_{12}+B^2\tau_{22},\end{aligned}$$

and the coefficients of $\sqrt{-1}$ in the quadratic forms $\tau(0,1)^2$ and $\tau(2A,B)^2$ are positive (p. 176).

Since the canonical forms involve the invariant alone, it follows that *any two singular relations with the same invariant can be linearly transformed into each other.* This important theorem shows that in the elliptic case a linear transformation can be found which reduces the singular relation to the form

$$k\tau_{12}-1=0,$$

for this has invariant $\Delta=k^2$. We shall suppose that this has been done; then different elliptic surfaces are distinguished by different values of the positive integer k.

§ 113. PARAMETRIC CURVES.

The chief geometrical peculiarity of elliptic Kummer surfaces† is the existence upon them of two families of curves, $u_1=$ const. and $u_2=$ const., which, since the coordinates of a point on any one curve are expressible in terms of elliptic functions of one parameter, are algebraic and of deficiency 1.

* Humbert, *Liouville*, sér. 5, v, 245.
† Humbert, *Amer. Jour.*, XVI, 221.

On elliptic surfaces, as on the general surface, the point (a_1, a_2) is the same as the point $(\pm a_1 + P_1, \pm a_2 + P_2)$, where (P_1, P_2) is any pair of periods; hence the curves

$$u_1 = a_1,$$
$$u_2 = a_2,$$

intersect in all the different points given by

$$u_1 = \pm a_1 + P_1,$$
$$u_2 = \pm a_2 + Q_2,$$

where (Q_1, Q_2) is any pair of periods, and the ambiguous signs are independent. These are the same as the points given by

$$u_1 = \pm a_1 + P_1 - Q_1,$$
$$u_2 = a_2.$$

Since $(1, 0)$ and $(k\tau_{11}, k\tau_{12} - 1)$ are pairs of periods, whatever integer k may be, and in the elliptic case when $k\tau_{12} = 1$ the latter is $(k\tau_{11}, 0)$, it follows that all the distinct points of intersection are given by

$$u_1 = \pm a_1 + m\tau_{11} + n\tau_{12},$$
$$u_2 = a_2,$$

where $0 \leqslant m \leqslant k - 1$, $0 \leqslant n \leqslant k - 1$. Hence there are $2k^2$ common points.

We see that the coördinates are doubly periodic functions of u_1 alone, the periods being 1 and $k\tau_{11}$; in fact if Θ is a double theta function of the second order and zero characteristic, and if in the relation

$$\Theta(u_1 + \tau_{11}\alpha_1 + \tau_{12}\alpha_2 + \beta_1,\ u_2 + \tau_{21}\alpha_1 + \tau_{22}\alpha_2 + \beta_2)$$
$$= \Theta(u_1, u_2) \exp\{-4\pi i(\alpha_1 u_1 + \alpha_2 u_2) - 2\pi i(\tau_{11}\alpha_1^2 + 2\tau_{12}\alpha_1\alpha_2 + \tau_{22}\alpha_2^2)\},$$

we put $\qquad \alpha_1 = k\bar{\alpha}, \quad k\tau_{11} = \tau_1, \quad \alpha_2 = 0, \quad \beta_2 = -\bar{\alpha},$

we find

$$\Theta(u_1 + \tau_1\bar{\alpha} + \beta_1, u_2) = \Theta(u_1, u_2) \exp(-4k\pi i\bar{\alpha}u_1 - 2k\pi i\tau\bar{\alpha}^2),$$

showing that Θ is a single theta function of u_1 of order $2k$. The number of zeros* not differing by multiples of 1 and τ_1 is $2k$, and a parameter curve $u_2 =$ const. cuts one of the coordinate planes in $2k$ points, and is therefore of order $2k$.

If a surface S of order k be made to pass through $2k^2$ arbitrary points of the curve $u_1 = a_1$, it will contain it entirely, since its deficiency is 1. If, further, S be made to pass through an arbitrary point of $u_2 = a_2$, it will cut this curve in $2k^2 + 1$ points

* Krazer, *Lehrbuch der Thetafunktionen*, p. 41.

and therefore pass through it. Now $2k^2+1$ is exactly the number of conditions which the complete intersection of Kummer's surface with an undetermined k-ic surface can be made to satisfy, and so we have the theorem that the curves $u_1=a_1$ and $u_2=a_2$ form together the complete intersection with a surface of order k.

Since the curve $u_2=a_2$ was determined by a single point of it, the parametric curves belong to two linear systems of dimension 1, and either system is cut out by k-ic surfaces through any curve of the other system.

Analytical proofs of these theorems may be obtained by using *single theta functions* defined by

$$\vartheta_{\alpha\beta}(u,\tau)=\Sigma\exp\{2\pi i(n+\tfrac{1}{2}\alpha)(n+\tfrac{1}{2}\beta)+\pi i\tau(n+\tfrac{1}{2}\alpha)^2\},$$

where all the letters represent single quantities and the summation is for all integer values of n.

It may be shown directly, by rearranging the terms, that the double theta functions to which the coordinates are equated may be replaced by the expressions

$$\sum\sum_{\nu_1,\nu_2=0}^{k-1}\exp\{-4\pi i k^{-1}(\nu_1+\tfrac{1}{2}\alpha_2)(\nu_2+\tfrac{1}{2}\alpha_1)\}$$
$$\vartheta_{\alpha_1 0}\{2u_1+(2\nu_1+\alpha_2)k^{-1},2\tau_{11}\}\,\vartheta_{\alpha_2 0}\{2u_2+(2\nu_2+\alpha_1)k^{-1},2\tau_{22}\},$$

where $(\alpha_1,\alpha_2)=(10),(11),(01),(00)$ in succession.

The two parametric curves $u_1=a_1$, and $u_2=a_2$, are given by the equation

$$\vartheta_{11}(ku_1-ka_1,k\tau_{11})\,\vartheta_{11}(ku_1+ka_1,k\tau_{11})$$
$$\vartheta_{11}(ku_2-ka_2,k\tau_{22})\,\vartheta_{11}(ku_2+ka_2,k\tau_{22})=0.$$

The left side is a double theta function of order $2k$ and zero characteristic, and therefore represents the complete intersection with a surface of order k.

§ 114. UNICURSAL CURVES.

The curves $u_1=a_1+\tfrac{1}{2}P_1$ and $u_1=-a_1+\tfrac{1}{2}P_1$ are the same, (P_1,P_2) being any pair of periods. When a_1 is small the points $(u_1+\tfrac{1}{2}P_1,u_2)$ and $(-a_1+\tfrac{1}{2}P_1,u_2)$ are near. Hence the curve $u_1=\tfrac{1}{2}P_1$ occurs *repeated* in the linear system and is therefore of order k. Again the points $(a_1+\tfrac{1}{2}P_1,u_2)$ and $(-a_1+\tfrac{1}{2}P_1,-u_2+P_2)$ are the same, so that at any point of the curve $u_1=\tfrac{1}{2}P_1$ the variable parameter u_2 can have two values whose sum is P_2; the coordinates are therefore proportional to even doubly periodic functions of a new variable $u_2-\tfrac{1}{2}P_2$, and hence are rationally expressible in terms of one even elliptic function. Hence *the curves obtained by equating the parameters to half-periods are of order k and unicursal.*

When k is odd, there lie on the curve

$$u_1 = \tfrac{1}{2}(\tau_{11}\alpha_1 + \tau_{12}\alpha_2 + \beta_1),$$

the nodes associated with the characteristics

$$\begin{pmatrix}\alpha_1 & \alpha_2 \\ \beta_1 & 0\end{pmatrix} \begin{pmatrix}\alpha_1 & \alpha_2 \\ \beta_1 & 1\end{pmatrix} \begin{pmatrix}\alpha_1 & \alpha_2+1 \\ \beta_1+1 & 0\end{pmatrix} \begin{pmatrix}\alpha_1 & \alpha_2+1 \\ \beta_1+1 & 1\end{pmatrix},$$

for the first parameters of these nodes are either equal to the preceding value of u_1 or differ from it by $\frac{1}{2}(\tau_{12}+1)$, which is here equal to the period $\frac{1}{2}(k+1)\tau_{12}$, on account of the singular relation. The sum of these four characteristics is zero and the parity of their product is odd: accordingly they belong to a Rosenhain tetrad. Hence there are only four unicursal curves in the first family of parameter curves, and each passes through a different Rosenhain tetrad of nodes.

Similarly the curve

$$u_2 = \tfrac{1}{2}(\tau_{21}\alpha_1 + \tau_{22}\alpha_2 + \beta_2)$$

passes through the nodes associated with the characteristics

$$\begin{pmatrix}\alpha_1 & \alpha_2 \\ 0 & \beta_2\end{pmatrix} \begin{pmatrix}\alpha_1 & \alpha_2 \\ 1 & \beta_2\end{pmatrix} \begin{pmatrix}\alpha_1+1 & \alpha_2 \\ 0 & \beta_2+1\end{pmatrix} \begin{pmatrix}\alpha_1+1 & \alpha_2 \\ 1 & \beta_2+1\end{pmatrix},$$

forming a Rosenhain tetrad, having one corner in common with the preceding one. Hence the unicursal curves of the second family pass through the tetrads of another group-set and one curve of each family passes through each node. Since a unicursal curve counted twice is a particular case of a parameter curve, every intersection of two unicursal curves, except their common node, counts as four common points of two parameter curves, the node itself counting as two, so that the two unicursal curves cut in one node and $\frac{1}{2}(k^2-1)$ other points.

A surface S of order $\frac{1}{2}(k+1)$ can be drawn through these $\frac{1}{2}(k^2+1)$ common points and $\frac{1}{2}(k+1)$ arbitrary points on each curve, for this makes $\frac{1}{2}(k+1)^2+1$ conditions. S then passes through both curves. Now each curve meets the common face of the two Rosenhain tetrahedra in three nodes and $\frac{1}{2}(k-3)$ points of contact, so that S meets the conic in that face in $k+3$ points, and therefore contains it entirely. Hence two unicursal curves of different systems, together with one conic, form the complete intersection with a surface of order $\frac{1}{2}(k+1)$.

When k is even the curve

$$u_1 = \tfrac{1}{2}(\tau_{11}\alpha_1 + \tau_{12}\alpha_2 + \beta_1)$$

passes through the nodes associated with the characteristics

$$\begin{pmatrix}\alpha_1 & \alpha_2\\ 0 & 0\end{pmatrix} \quad \begin{pmatrix}\alpha_1 & \alpha_2\\ 0 & 1\end{pmatrix} \quad \begin{pmatrix}\alpha_1 & \alpha_2\\ 1 & 1\end{pmatrix} \quad \begin{pmatrix}\alpha_1 & \alpha_2\\ 1 & 0\end{pmatrix},$$

forming a Göpel tetrad. Similarly the curve

$$u_2 = \tfrac{1}{2}(\tau_{21}\alpha_1 + \tau_{22}\alpha_2 + \beta_2)$$

passes through the same nodes. There are four unicursal curves in each system of parameter curves, and they intersect by pairs in the nodes of four Göpel tetrads of a group-set. Two unicursal curves through the same nodes cut again in $\frac{1}{2}k^2 - 2$ points. A surface of order $\frac{1}{2}k$ can be drawn through $\frac{1}{2}k^2 + 1$ of these common points and then contains both curves.

§ 115. GEOMETRICAL INTERPRETATION OF THE SINGULAR RELATION $k\tau_{12} = 1$.

Each pencil of parametric curves determines an *involution* on each singular conic, that is, groups of k points depending linearly on one variable, so that each group is determined by any one point of the group. Four of these groups are cut out by unicursal curves (counted twice), and are of a special character. Upon this depends the situation of the six nodes on each conic and the geometrical interpretation of the singular relation.

When k is even the four unicursal curves pass through Göpel tetrads of nodes, forming a group-set, and three of the tetrads have each two nodes in common with any trope, as is at once seen from

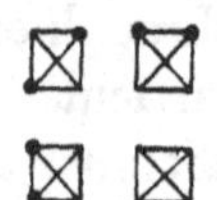

the diagram, in which the four tetrads and one set of coplanar nodes are indicated.

Hence three groups of the involution consist of two nodes and $\frac{1}{2}(k-2)$ points counted twice, and one group consists of $\frac{1}{2}k$ points counted twice. The other pencil of parametric curves determines the same involution.

When k is odd the unicursal curves pass through Rosenhain tetrads of which one contains three nodes of a given trope and the

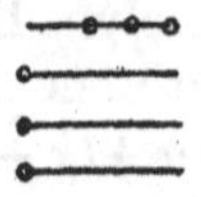

others one each. Hence one group of the involution consists of three nodes and $\frac{1}{2}(k-3)$ other points counted twice, and three groups consist of one node and $\frac{1}{2}(k-1)$ points counted twice (pairs of coincident points). The other pencil of parameter curves determines another involution in which the parts played by the two sets of three nodes are interchanged.

To illustrate this, consider the case when $k=2$.

The nodes on each conic form three pairs of an involution, and therefore the chords joining them are concurrent. This has been shown to be the condition for a *tetrahedroid*. There exist eight unicursal curves of order 2, that is conics, intersecting by pairs in four Göpel tetrads of nodes; these tetrads are therefore coplanar. Three of these pairs of conics cut any singular conic in three pairs of nodes; the remaining two conics touch the singular conic at the double points of the involution to which the nodes belong.

There exist on the surface two pencils of elliptic quartic curves (intersections of pairs of quadric surfaces), obtained by making one parameter or the other constant. These results agree with what has been proved before by other methods.

Next suppose $k=3$.

Each pencil of parameter curves consists of sextics* cutting each singular conic in groups of three points of an involution including one group of three nodes, say 1, 3, 5. Now the chords joining points of the same group touch a conic C, since two pass through any point of the singular conic. C touches the sides of the triangle 135, and since the two tangents to it from each of the other three nodes 2, 4, 6, are coincident, C passes through them. Hence a necessary geometrical condition for an elliptic surface of invariant 9 is that *a conic should pass through three nodes and touch the lines joining the other three*; it is easy to show that this condition is sufficient. The corresponding relation among the coefficients k_s, or *modular equation*, is

$$|(k_{2r-1}-k_{2s})^{-\frac{1}{2}}|=0, \qquad (r, s=1, 2, 3)$$

and the symmetry of this result, as well as the existence of the other pencil of parametric curves, shows that another conic can be circumscribed about 1, 3, 5, and inscribed in 2, 4, 6.

The group of nodes 1, 3, 5 of the first involution is cut out on the singular conic by a twisted cubic passing through the remaining node 135 of the Rosenhain tetrad determined by them. This cubic is projected from the node 135 into the conic circum-

* Further properties are given by Humbert, *Amer. Jour.* XVI, 249.

scribing 1, 3, 5, and inscribed in 2, 4, 6. Similarly there is a twisted cubic of the other system passing through the nodes 2, 4, 6, 246, and is projected from the last, which is the same as 135, into the conic C. The two cubics, taken together, form a degenerate member of the ordinary family of sextics through six nodes cut out by quadrics through one conic.

§ 116. INTERMEDIARY FUNCTIONS.

Consider now Kummer's surface defined by theta functions in the usual way, except that the periods are connected by a singular relation whose invariant is not a square. They are characterised geometrically by the breaking up of some of the ordinary curves lying on the surface into curves of lower order which do not in general exist.

The general transcendental theory shows that the hyperelliptic equation of any curve is $F(u)=0$, where

$$\log\{F(u+\tau\alpha+\beta)/F(u)\}$$

is linear in u_1, u_2, 1. By multiplying F by the exponential of a quadratic in u_1, u_2, 1, and making the periodic conditions consistent we can arrange that

$$F(u_1+1, u_2)=F(u),$$

$$F(u_1, u_2+1)=F(u)\exp 2\pi i m u_1,$$

$$F(u_1+\tau_{11}, u_2+\tau_{21})$$
$$=F(u)\exp\{-2\pi i(n_{11}u_1+\overline{n_{12}+m\tau_{11}}\,.\,u_2)+\text{const.}\},$$

$$F(u_1+\tau_{12}, u_2+\tau_{22})$$
$$=F(u)\exp\{-2\pi i(n_{21}u_1+\overline{n_{22}+m\tau_{12}}\,.\,u_2)+\text{const.}\},$$

where m, n_{11}, n_{12}, n_{21}, n_{22} are integers and the periods are connected by the relation

$$n_{11}\tau_{12}+(n_{12}+m\tau_{11})\tau_{22}=n_{21}\tau_{11}+(n_{22}+m\tau_{12})\tau_{12}+\text{integer}.$$

In the ordinary case all the integer coefficients of

$$\tau_{11}, \tau_{12}, \tau_{22}, \tau_{11}\tau_{22}-\tau_{12}^2, \text{ and } 1,$$

in this relation must be zero, and then F becomes a theta function of order n_{11} or n_{22}; but when a singular relation exists, the preceding relation may be made equivalent to it and then functions different from theta functions can exist and have the preceding properties. By a linear transformation it is possible to make $m=0$ and reduce the singular relation to the form

$$n_{21}\tau_{11}+(n_{22}-n_{11})\tau_{12}-n_{12}\tau_{22}=0,$$

which includes the two canonical forms for odd and even invariants.

Take new arguments (U) defined by

$$|n|\,U_1 = n_{11}u_1 + n_{12}u_2,$$
$$|n|\,U_2 = n_{21}u_1 + n_{22}u_2,$$

where $$|n| = n_{11}n_{22} - n_{12}n_{21};$$

in matrix notation the substitution is

$$|n|\,U = nu.$$

The new periods are the elements of a matrix τ which is defined by

$$|n|\,\tau = n\tau,$$

and is symmetrical on account of the singular relation. Then

$$F(u) = \Phi(U),$$

where $$\Phi(U_1+1, U_2) = \Phi(U_1, U_2+1) = \Phi(U),$$
$$\Phi(U_1+\tau_{11}, U_2+\tau_{21}) = \Phi(U)\exp\{-2\pi i\,|n|\,U_1 + \text{const.}\},$$
$$\Phi(U_1+\tau_{12}, U_2+\tau_{22}) = \Phi(U)\exp\{-2\pi i\,|n|\,U_2 + \text{const.}\},$$

showing that constants c_1, c_2, can be found so that $\Phi(U)$ is a theta function of the arguments $(U-c)$, of order $|n|$ and zero characteristic. We shall take (c) to be a half-period.

The integer elements of the matrix (n) satisfy three inequalities, as follows. Since the invariant of the singular relation

$$n_{21}\tau_{11} + (n_{22}-n_{11})\,\tau_{12} - n_{12}\tau_{22} = 0$$

is positive, we have

$$(n_{11}+n_{22})^2 - 4\,(n_{11}n_{22} - n_{12}n_{21}) > 0.$$

Again, in order that the matrix τ may be suitable for the construction of a theta function, if $\tau_{rs} = \tau_{rs}' + i\tau_{rs}''$,

then $$|\tau''| > 0,$$
but $$|n|^2\,|\tau''| = |n\tau''| = |n|\,|\tau''|,$$
and $$|\tau''| > 0,$$
so that $$|n| = n_{11}n_{22} - n_{12}n_{21} > 0.$$

Lastly, $$|n|\,\tau_{11}'' = n_{11}\tau_{11}'' + n_{12}\tau_{21}'' > 0,$$

and it may be proved by elementary considerations that this inequality may be replaced by

$$n_{11} + n_{22} > 0.$$

These three inequalities are implied by the single condition

$$n_{11} + n_{22} > 2\sqrt{n_{11}n_{22} - n_{12}n_{21}}.$$

§ 117. SINGULAR CURVES.

Taking $\Phi(U;\tau)$ to be a theta function with definite parity and characteristic, we find that a family of *singular curves* exists whose equation is $F(u)=0$ where

$$F(u+\tau\bar{a}+\bar{\beta})=(-)^{a\bar{\beta}+\bar{a}\beta}F(u)\exp\{-\pi i\bar{a}n(2u+\tau\bar{a})\},$$

and F is either even or odd.

This differs from the periodic property of a theta function by the presence of the matrix (n) in place of a single integer. If we assume an expansion

$$F(u)=e^{\pi iau}\sum A_s\,e^{2\pi isu}$$

the preceding relation gives $A_{s_1+n_{11},\,s_2+n_{12}}$, and $A_{s_1+n_{21},\,s_2+n_{22}}$ in terms of $A_{s_1,\,s_2}$; whence it follows that the number of linearly independent functions $F(u)$ having the same characteristic $a\beta$ and matrix n is the area of the parallelogram whose corners, referred to rectangular axes, are $(0, 0)$, (n_{11}, n_{12}), (n_{21}, n_{22}), $(n_{11}+n_{21}, n_{12}+n_{22})$, that is, $|n|$.

The nodes through which the singular curve $F(u)=0$ passes are given by the half-periods for which F vanishes. These are seen from the periodic relation to be $\frac{1}{2}(\tau\bar{a}+\bar{\beta})$ where

$$\bar{a}\beta+a\bar{\beta}+\bar{a}n\bar{\beta}$$

has the opposite parity to F. Hence all the curves of one family pass through the same nodes.

If the elements of the matrix

$$n=\begin{pmatrix} n_{11} & n_{12}\\ n_{21} & n_{22}\end{pmatrix}$$

satisfy the condition

$$n_{11}+n_{22}>\sqrt{n_{11}n_{22}-n_{12}n_{21}},$$

so also do the elements of

$$|n|\,n^{-1}=\begin{pmatrix} n_{22} & -n_{12}\\ -n_{21} & n_{11}\end{pmatrix}$$

and the singular relations are the same in both cases.

By adding the exponents in the period relations we see that the product of two intermediary functions, one from each family, is a theta function of order $n_{11}+n_{22}$ and characteristic the sum of those of its factors.

Hence the curve $F(u)=0$ is of order $n_{11}+n_{22}$ and arises from the breaking up of an ordinary curve of order $2(n_{11}+n_{22})$. The base points of this ordinary family are divided among the two singular curves which intersect in a certain number of other nodes

It can be proved that the number of intersections of two singular curves, distinguished by different matrices n and n', is

$$\tfrac{1}{2}(n_{11}n_{22}' + n_{22}n_{11}' - n_{12}n_{21}' - n_{21}n_{12}').$$

On putting $n_{11}' = n_{22}' = 2$, $n_{12}' = n_{21}' = 0$, the second curve becomes a plane section, and we get the order $n_{11} + n_{22}$.

§ 118. SINGULAR SURFACES WITH INVARIANT 5.

The existence of singular curves upon a singular surface of invariant 5 depends upon the choice of four integers $n_{11}, n_{12}, n_{21}, n_{22}$ satisfying

$$n_{11} + n_{22} = \sqrt{4(n_{11}n_{22} - n_{12}n_{21}) + 5}.$$

The smallest values give the most interesting result and accordingly we take $n_{11} = n_{12} = n_{21} = 1$, $n_{22} = 2$. The corresponding equation $F(u) = 0$ represents a curve of order $n_{11} + n_{22}, = 3$, and if the characteristic is zero, it passes through the nodes $\tfrac{1}{2}(\tau\alpha + \beta)$ where

$$\alpha_1\beta_1 + \alpha_1\beta_2 + \alpha_2\beta_1 + 2\alpha_2\beta_2$$

has the opposite parity to F. Now it is easily found that the congruence

$$(\alpha_1 + \alpha_2)\beta_1 + \alpha_1\beta_2 \equiv 0 \pmod{2}$$

has ten solutions, and the congruence

$$(\alpha_1 + \alpha_2)\beta_1 + \alpha_1\beta_2 \equiv 1 \pmod{2}$$

has six solutions; the latter are underlined in the diagram

$$\begin{array}{cccc} dd & \underline{ac} & \underline{ba} & \underline{cb} \\ ab & da & \underline{cc} & bd \\ bc & cd & db & \underline{aa} \\ ca & \underline{bb} & ad & dc \end{array}$$

and are seen to form a Weber-hexad. The cubic cannot pass through the ten nodes since five of them lie in the trope (ab), and so must pass through the hexad. It is projected from the node (aa) by a quadric cone passing through the intersections of the tropes (cd), (dc), (bd), (db), (cb), (cd) taken consecutively and therefore touching the remaining trope (bc) which passes through the node (aa). The fact that such a conic can be drawn is a consequence of the singular relation, and it is easy to express the conditions in terms of the constants of the surface.

The values $n_{11}=2$, $n_{22}=1$, $n_{12}=n_{21}=-1$, give another cubic passing through the nodes $\frac{1}{2}(\tau\alpha+\beta)$ where

$$2\alpha_1\beta_1-\alpha_1\beta_2-\alpha_2\beta_1+\alpha_2\beta_2\equiv 1 \text{ (mod. 2)}$$

or
$$(\alpha_1+\alpha_2)\beta_2+\alpha_2\beta_1\equiv 1 \text{ (mod. 2)},$$

that is, the underlined hexad in the diagram

$$\begin{array}{cccc} dd & ac & ba & cb \\ \underline{ab} & da & \underline{cc} & bd \\ \underline{bc} & cd & db & \underline{aa} \\ \underline{ca} & \underline{bb} & ad & dc \end{array}$$

Thus the two cubics taken together form a sextic of the family passing through the six nodes in the trope (dd); they both pass through the same three additional nodes (aa), (bb), (cc).

If $\alpha'\beta'$ is the characteristic of $F(u)$ the singular curve in the first case passes through the nodes $(\alpha\beta)$ given by

$$\alpha_1\beta_1'+\alpha_2\beta_2'+\alpha_1'\beta_1+\alpha_2'\beta_2+\alpha_1\beta_1+\alpha_1\beta_2+\alpha_2\beta_1\equiv 0 \text{ or } 1,$$

or
$$(\alpha_1+\alpha_2+\alpha_1')(\beta_1+\beta_2')+(\alpha_1+\alpha_2')(\beta_2+\beta_1')\equiv 0 \text{ or } 1,$$

so we have only to add the characteristic

$$\begin{pmatrix}\alpha_2' & \alpha_1'-\alpha_2' \\ \beta_2' & \beta_1'\end{pmatrix}$$

to the solutions of the former congruence. Hence corresponding to the matrix

$$(n)=\begin{pmatrix}1 & 1 \\ 1 & 2\end{pmatrix}$$

there are sixteen cubics on the surface, forming a group-set; of these six pass through each node. A similar result holds for the cubics associated with the matrix

$$(n)=\begin{pmatrix}2 & -1 \\ -1 & 1\end{pmatrix}.$$

Since $|n|=1$ each singular family contains only one curve.

§ 119. SINGULAR SURFACES WITH INVARIANT 8.

The elements of the matrix (n) have to satisfy

$$n_{11}+n_{22}=\sqrt{4(n_{11}n_{22}-n_{12}n_{21})+8}.$$

The smallest value of $|n|$ is 2 giving $n_{11}+n_{22}=4$. We find therefore singly infinite families of quartics, taking

$$\begin{pmatrix}n_{11} & n_{12} \\ n_{21} & n_{22}\end{pmatrix}=\begin{pmatrix}2 & 1 \\ 2 & 2\end{pmatrix},$$

the singular relation has the canonical form

$$-2\tau_{11} + \tau_{22} = 0.$$

The singular family of characteristic $\alpha'\beta'$ passes through the nodes $(\alpha\beta)$ given by

$$\alpha\beta' + \alpha'\beta + \alpha n\beta \equiv 0 \text{ or } 1 \pmod{2},$$

that is $\quad (\alpha_1 + \alpha_2')(\beta_2 + \beta_1') + \alpha_2\beta_2' + \alpha_1'\beta_1 \equiv 0$ or 1.

First take $\quad \alpha_1' = \alpha_2' = \beta_1' = \beta_2' = 0;$

the congruence $\quad \alpha_1\beta_2 \equiv 1$

has one solution $\alpha_1 = \beta_2 = 1$, giving four nodes since α_2 and β_1 are arbitrary. These nodes are

$$\begin{pmatrix}1 & 0\\0 & 1\end{pmatrix}\ \begin{pmatrix}1 & 1\\0 & 1\end{pmatrix}\ \begin{pmatrix}1 & 0\\1 & 1\end{pmatrix}\ \begin{pmatrix}1 & 1\\1 & 1\end{pmatrix},$$

or $\quad (aa) \quad (ba) \quad (ab) \quad (bb)$

and form a Göpel tetrad. The alternative congruence

$$\alpha_1\beta_2 \equiv 0$$

gives the remaining twelve nodes, which are inadmissible since there are tropes containing six of them.

Hence there exists on this surface a singular family of quartics passing through four nodes and depending on one parameter. This parameter can be chosen to make the quartic pass through an additional node which will then be a double point on the curve since it is not a base node. By projecting from this node we get a quadric cone passing through the four lines of intersection of four tropes taken in order, and touching the remaining two. The reciprocal property holds for the six nodes in any trope and characterises the surface: *a conic can be described through two nodes to touch the sides of a quadrilateral formed by the other four.*

By taking different values for α_2' and β_1' we get three other families of quartics passing through the remaining three tetrads of the group-set. Again taking

$$\begin{pmatrix}n_{11} & n_{12}\\n_{21} & n_{22}\end{pmatrix} = \begin{pmatrix}2 & -2\\-1 & 2\end{pmatrix}$$

we find a family of quartics of characteristic $\alpha'\beta'$ passing through the nodes given by

$$\alpha_1\beta_1' + \alpha_2\beta_2' + \alpha_1'\beta_1 + \alpha_2'\beta_2 + \alpha_2\beta_1 \equiv 0 \text{ or } 1,$$

so that it is necessary only to interchange α and β, α' and β'. Thus there is a second family of quartics through the nodes (aa), (ab), (ba), (bb), and by the general theory any two curves, one from each family, lie on a quadric.

§ 120. BIRATIONAL TRANSFORMATIONS OF KUMMER SURFACES INTO THEMSELVES.

Examples of these are the fifteen collineations which with identity make up the group of sixteen linear transformations upon which the whole theory of the general Kummer surface depends. Other transformations, not linear, are geometrically evident, namely the sixteen projections of the surface upon itself from the nodes, and the correlative transformations by means of tangent planes collinear with a trope. The question was proposed by Klein*, and answered by Humbert†, as to whether any other such transformations exist.

A method is given in § 96 for finding in succession *all* the quartic surfaces into which the *general* Kummer surface can be transformed birationally; among these, for different orders of the transformation, the surface itself occurs, as for example when $n=6, s=0$. Two examples are pointed out by Hutchinson‡ who refers the surface to a Göpel tetrad of nodes. The equation has the form

$$A(x^2t^2+y^2z^2)+B(y^2t^2+z^2x^2)+C(z^2t^2+x^2y^2)+Dxyzt$$
$$+F(yt+zx)(zt+xy)+G(zt+xy)(xt+yz)$$
$$+H(xt+yz)(yt+zx)=0$$

and is unchanged by the transformation§

$$x'x=y'y=z'z=t't,$$

so that by using different tetrads we obtain in this way an infinite group of birational transformations. Again the equation of Weddle's surface

$$|\,x_s^{-1},\, x_s,\, e_s,\, f_s\,|=0 \qquad (s=1, 2, 3, 4)$$

is unchanged by the same substitution. There are fifteen equations of this form, referred to different tetrahedra of nodes, and we obtain in this way another infinite group of transformations of Kummer's surface since there is a one-one relation between the two surfaces.

There exist special Kummer surfaces which admit other transformations than those indicated. We consider only the problem of finding those surfaces which can be *linearly* transformed into themselves|| otherwise than by the group of sixteen collineations.

* Klein, *Math. Ann.* (1885), XXVII, 142.

† Humbert, *Liouville* (1893), sér. 4, IX, 465.

‡ Hutchinson, *Amer. Bull.* (1901), VII, 211.

§ Showing that the six parameters of three nodes on a unicursal quartic are those of six nodes on a conic, Humbert, *Comptes Rendus* (1901), CXXXIII, 425.

|| Kantor, *Amer. Jour.* (1897), XIX, 86.

Suppose that such a transformation exists; it changes the nodes on one conic into the nodes on the same or some other conic, and by combining one of the sixteen collineations we may arrange that the conic is the same. Conversely if the six nodes 1, 2, 3, 4, 5, 6 in the plane 0 are projectively related to the same nodes in a different order 1′, 2′, 3′, 4′, 5′, 6′, then it is easily seen that a linear transformation can be found which interchanges the Göpel tetrahedra of tropes 0, 12, 34, 56 and 0, 1′2′, 3′4′, 5′6′, and leaves the surface unaltered. Thus the problem is reduced to finding the conditions under which six points on a conic can be linearly transformed into themselves*.

Now in a linear transformation of a conic into itself the chords joining corresponding points touch another conic having double contact at the self-corresponding points.

These conics may be projected into concentric circles and then any cyclic permutation must represent a regular polygon inscribed in one circle and circumscribed about the other; for instance a cycle (12) means that the points 1 and 2 are the ends of a diameter, and the inner circle has zero radius. Now any permutation can be arranged as a set of cycles, the elements of each cycle being permuted in cyclic order among themselves; in the present case there must not be more than two points unchanged, and the remaining points must be the corners of one or more regular polygons, the number of corners being the same for different polygons. Under these restrictions the only possible permutations are represented by the following six types:

I.	(12) (34) (56)	three diameters	single tetrahedroid
II.	(12) (34) (5) (6)	two diameters, I	double tetrahedroid
III.	(123) (456)	two triangles	triple tetrahedroid
IV.	(123456)	hexagon	quadruple tetrahedroid
V.	(12345) (6)	pentagon and I	singular, invariant 5
VI.	(1234) (5) (6)	square, I, J	sextuple tetrahedroid

* Bolza, *Amer. Jour.*, x, 47.

The third column describes the projective nature of the situation of coplanar nodes regarded as lying on a circle passing through I and J the circular points at infinity. By comparison with the results of § 57 we identify the corresponding surface as a tetrahedroid except in case V.

It is interesting to see how these surfaces arise in the theory of the birational transformation of the *hyperelliptic* surface into itself*. The general hyperelliptic surface is defined by equating the coordinates to theta functions of the same order and zero characteristic, and in general each point has a *single* pair of parameters u_1, u_2. Then du_1 and du_2 are differentials "of the first kind" for the surface and must be linearly transformed when the surface undergoes a birational transformation. Thus we are led to the special kind of transformation of theta functions known as *complex multiplication* in which the new periods are the same as the old, and in order to be uniquely reversible the transformation must be of the first order. There are two cases, according as the transformation is ordinary or singular (pp. 202, 203). In the former case it can be shown that when the surface is not singular the only transformations are given by

$$u_1' = u_1 + \text{const.}, \qquad u_2' = u_2 + \text{const.},$$

and by $$u_1' = -u_1 + \text{const.}, \quad u_2' = -u_2 + \text{const.};$$

Kummer's surface is distinguished from the general hyperelliptic surface by the fact that each point has *two* pairs of parameters (u_1, u_2) and $(-u_1, -u_2)$, and the preceding reasoning fails. In fact the transformation

$$u_1' = u_1 + a_1, \quad u_2' = u_2 + a_2$$

gives two distinct points $(u + a)$ and $(-u + a)$ corresponding to one point (u) unless (a) is a half-period, in which case the transformation is one of the group of sixteen collineations. The preceding equations do, however, express a one-one relation between the points of the tangent sections (v) and $(v + a)$, where (v) is arbitrary, because each is a univocal curve (p. 190), and this affords a proof that all tangent sections have the same moduli.

When the hyperelliptic surface is singular it may admit other birational transformations depending on ordinary transformations of theta functions. It can be shown that this depends

* Humbert, *Liouville*, sér. 5, VI, 367.

on a linear transformation of the integrals $\int dx/y$ and $\int x\,dx/y$ where

$$y^2 = (x-k_1)(x-k_2)(x-k_3)(x-k_4)(x-k_5)(x-k_6),$$

and this again depends on a linear transformation of the sextic into itself, leading to the same six sets of relations among the constants, to each of which corresponds a certain type of singular Kummer surface.

INDEX.

CAMBRIDGE: PRINTED BY JOHN CLAY, M.A. AT THE UNIVERSITY PRESS.

Printed in the United States
By Bookmasters